BLOODY HELL

BLOODY HELL

The Price Soldiers Pay

Dan Hallock
Foreword by Simon Weston

The Plough Publishing House

©1999 by The Plough Publishing House
of the Bruderhof Foundation

Farmington, PA 15437 U.S.A.
Robertsbridge, E. Sussex, TN32 5DR U.K.

All Right Reserved

First Printing June 1999
Second Printing July 1999
Third Printing July 1999

Cover photograph ©The Defence Picture Library

A catalogue record for this book
is available from the British Library.

ISBN 0-87486-969-2

Printed in the UK

Contents

Foreword

SOMETIMES I THINK I NEVER went to war – war came to me. It hit, burned, changed, and tempered me. And taught me many things.

Bad people, I learned, don't go to war. It's the young who go to war, the nice lads. And it's the civilians who become the real innocent casualties of war. But the people who actually wage war are so far behind the lines that they don't even get a smell of cordite, let alone hear the shells explode.

That's part of the reality of conflict. The other part is that, when you go to war, it becomes your job to kill people just as nice as you.

The Argentinian flying the A-4 Skyhawk that bombed the landing craft *Sir Galahad* in "Bomb Alley" was a guy named Carlos. He's a nice guy. I know him; I've met him and his wife Gracía – they're lovely people. I have no quarrel with him. When he dropped the bomb that killed so many of my friends and left me burned, he was just a man doing a job. It was 1982, and his country was at war with mine; he had a job to do. If roles had been reversed, I know I would have done the same.

Both of us have had our share of nightmares. He saw my picture in the Argentinian papers, and it haunts him to know he caused such pain. As he said to me, "Why do you have to be such a nice person?" It would have made life so much easier for him if I'd been the devil. But as it was, we were all nice guys.

I'm not interested in war anymore. I'm not interested in the reunions, getting together with the old mates and patting each other on the back, saying, "Damn, didn't we do a good job?" What, shoot and kill and bomb their side? Lovely. It's a job. If you somehow come out the other end the winner, you'll find there was nothing victorious or glorious gained in the conflict. The only winners are the financial houses, the arms industry, and the politicians who've used the system and current affairs to aid and abet their desire for power.

Then you've got two sets of soldiers, which are the losers and the losers. It's just a matter of who loses most heavily. There are no winners.

And afterwards…afterwards, an inescapable truth remains: you have to live with yourself. And if you can't live with yourself, then there's no point in being alive.

There was a time after the Falklands when I drank far too much, trying to hide the pain that was eating at me from the inside. I tried to hide from the grief I felt for the friends I'd lost, as well as my own very real grief for the person I knew I'd no longer be.

I had a huge grieving process to go through. Thankfully, I learned not to bottle it up, but it took a long time. Otherwise, I'm sure, it would have killed me. I've reconciled things within myself. I don't go around hating people. I can't hate Carlos, because I know he didn't deliberately try to kill me or Jimmy Weaver or Andrew Walker, or any of the boys. It's not a personal thing. The way I see it, I don't have a cross to bear; I count myself lucky to be alive.

I'm no soothsayer; I can't see into the future, and I definitely can't take back the past. All I can do is try – and, believe me, it can be very difficult – to live by the lessons life has taught me. I want to enjoy my life, and I want others to be able to enjoy theirs, which is why I put so much time into working with inner-city youth through my own charity, Weston Spirit. I can't enjoy life if I'm twisted up in hatred and bitterness. I've learned you can't afford to carry that kind of baggage with you. If you do, it will destroy you. Because without joy, life is desperate.

There will always be war. That's the truth, as I see it, even if it sounds defeatist. But that doesn't have to stop me believing in love. I much prefer teaching my children to ride bicycles than teaching them to fire guns.

I believe that the more people you can dissuade away from violence, the better life is going to be, the more constructive life will be. But this won't happen if you sit back and take no action at all. We aren't just visitors to this planet. We have to take part to make a difference. And that doesn't mean going out and changing the world, altering its axis. It simply means each of us has to make a difference – in our own lives, in our families. We have to leave some small mark.

Ultimately, each of us has to take responsibility for our own actions, and for our own lives. There is no place for the old arguments. No matter who we are or what we've been through – the lessons of life have to take us forward. This is the message that lies, like a pearl, at the heart of *Bloody Hell.*

Simon Weston
Cardiff, 1999

A true war story is never moral. It does not instruct, nor encourage virtue, nor suggest models of proper human behavior, nor restrain men from doing the things men have always done. If a story seems moral, do not believe it. If at the end of a war story you feel uplifted, or if you feel that some small bit of rectitude has been salvaged from the larger waste, then you have been made the victim of a very old and terrible lie. There is no rectitude whatsoever. There is no virtue. As a first rule of thumb, therefore, you can tell a true war story by its absolute and uncompromising allegiance to obscenity and evil... You can tell a true war story if it embarrasses you. If you don't care for obscenity, you don't care for the truth... You can tell a true war story by the way it never seems to end... And in the end, really, there's nothing much to say about a true war story except maybe "Oh."

Tim O'Brien, *The Things They Carried*

On the surface, Lee looks like any hired laborer. His face and hands are tanned and weather-beaten, and his trademark flannel shirt and dangling cigarette identify him unmistakably. He is a quiet man who loves horses but more often than not shies away from people. And if you take the time to listen, he'll tell you why:

May 15, 1969, was the day that would change the rest of my life. I had joined the Army in August 1968 and asked to be sent to Vietnam. On May 18, I was on my way. I left from Fort Lewis, Washington, an army base where we had last-minute training. We flew on a civilian airline, and it took us eighteen hours to get there – a very long flight. None of us on the flight knew anyone else.

We landed at Cam Ranh Bay air base and were sent right away to a replacement battalion where we stayed until it was time to go to our new units. It was 118 degrees Fahrenheit (48 °C) when we got to Nam...

I was sent to a construction company in Don Duong, a tiny village beneath a huge dam built by the Japanese during World War II. There were four hundred of us stationed at the base of the dam, just a mile away from the village. There were no other Americans within forty miles.

My first job was to build the living and working areas. This was a brand new unit, and so at first we lived in large drain pipes, dug half-underground with tons of sandbags on top to stop mortars and rockets.

Within a few weeks our hooches (living/fighting bunkers) were done, and I joined a unit driving tractor trailers. Our job was to build roads from the central highlands to the outlying villages south and west. I trucked all day, six days a week. We had Sunday afternoons off.

Nam is a beautiful and amazing country, and I saw a lot of it over the next two years: the white sand beaches, the jungles, the mountains; monks and monasteries; a gold Buddha on a hilltop; animals I had never heard of before. I also saw truly ugly things: death.

On my Sunday afternoons off I would either go swimming at the dam, or go to the village for pot. You could buy a "key" (kilo) for $40 and do a lot of partying, with a lot of people, for a long time. There was always a place to buy pot. Most villages had a barn where you could go and sample different kinds before buying.

One day I met a girl on the road near the village. She knew a bit of English and we talked awhile. Her name was Chi. We liked each other from the start, and it wasn't long before she began to move around with me when our trucks were on the road for more than a few days. I even began to sneak out of the compound just to see her. That was very difficult: I had to crawl through barbed wire many times, and I remember getting stoned by a group of boys as I left one village. I had some bruises after that night, but I never got caught.

I had a lot of fun with Chi. I learned a lot about the Vietnamese people from her – their ways and customs, even a bit of their language.

Our trucks came under fire many times as we moved through the country, and so did our compound. Charlie (the Vietcong) hit us a lot, just to let us know he was there. He would blow up a couple of trucks, or throw in a few mortar rounds and blow up a few buildings, but I think he liked us being there, because we built roads that they still use today...

Once, when our convoy was hauling equipment to a road-building site, we came upon an accident. Everyone in front of me drove around it and kept right on going, but I stopped to see what was going on, so everyone behind me stopped too. An officer in one of the jeeps ahead of me flew back to see what the holdup was. He was mad when he saw me standing over a girl lying on the side of the road: "We cannot stop a whole convoy because of a gook," he shouted. "We can't leave her here," I shouted back. I picked her up and put her on the back of his jeep. He was ranting and raving; he wanted me to get her off and leave her – but I couldn't. Her leg was really messed up. She had been on the back of a motorcycle, and the young man who'd been driving her was dead.

I finally convinced the second lieutenant to help me. He ordered a driver to take over my truck; we took her up the road to a village to see if anyone knew her. Someone did and took us to her home. They were Montagnards: they lived in the hills outside the village. They are like Vietnamese hillbillies, a very tough people. Before long a crowd had gathered, and then a doctor came. The second lieutenant was in a hurry to leave, but I wanted to see if she was all right, as she was going in and out of consciousness. Her papa-san came and thanked me over and over. He wanted us to stay, but the lieutenant insisted that we go, so we did. I don't know what ever happened to her. I was reprimanded for putting the convoy in danger by stopping, but I didn't care. Spending half an hour with the Montagnards on their home ground and seeing the thankfulness in their faces had made it worth it.

Six months after I got to Nam I was transferred to a transportation company sixty miles away. But an officer going through the hooches found a huge bag of pot outside my window, so I was sent on. I became a convoy driver, and I was sent anywhere at any time. I loved being in a truck, driving around the country. I saw so much. I had made a lot of friends with the guys in my first unit, but I knew no one in my new company, and I decided to keep it that way. Chi moved to a village a mile away, and we continued to see each other whenever we could.

I traveled on just about every road in Vietnam, and some that weren't even roads yet. I remember driving to a firebase that had just been built: I had to take a ten-ton tractor with a loaded forty-foot flatbed trailer up a footpath! Mountain on one side and a cliff on the other. My trailer wheels hung in midair around some of the turns. I was sent first, before any of the other big trucks – I guess it was getting around that I was crazy, because it seemed I was put on all the dangerous convoys. Or maybe they liked the way I drove, high all the time…

When I came to the village to see Chi one night, she told me she was pregnant. I was stunned. Me – the father of a child – in a country 10,000 miles from home! Wow, what do I do?

We kept seeing each other as much as possible. She grew bigger and bigger. I was going to be a dad, and I started to think about what to do. I paid less attention to the other drivers, and spent more time getting high and thinking. I made very few friends from then on. On my way out of one base a guy ran up to me and asked me for a lift. He jumped in, and I just drove on. I paid no attention to him, but I could tell he was a new arrival. About fifteen miles later we were driving on a road through the rice paddies when all of a sudden I was lying in a ditch, looking up at my tractor. It was totally engulfed in flames, and my passenger was still sitting in it. He was screaming, screaming like I had never heard anyone scream before. The smell, the heat, the noise – I was confused for just a second. Then I realized we had been hit by rockets. I had been blown out of the truck. It burned as I lay in the water-filled ditch. And so did he, a soldier whose name I had never asked.

I became even harder toward people – all people – except Chi. I remember my commanding officer dragging me into his office after that, to make me write a letter home. I had become so hard inside that I hadn't written for almost two months. I wrote Mom a letter, but I didn't say much about what was going on in my head. Why was he killed, and not me? Why hadn't I at least asked him his name? This truly haunted me, but I never said a word about it to Chi. She knew something had happened, that I was somehow different, but she never asked. Why did he die, and not me?

Chi was getting bigger and bigger. I was on the road more and more. I volunteered for every convoy I could. Convoys don't move after dark much, it's too dangerous. But every time a truck had to go out at night I made sure I was there. I just wanted to be out of the compound; I hated sitting around. No one wanted to ride with me once they knew me. I didn't talk; I just drove and got high, which left the shotgun talking to himself…

Once I was sent out to retrieve two trucks that another convoy had left behind with mechanical troubles. We were going to tow them back. It was just before dark when we left – two wreckers and a jeep. I was driving a wrecker. We had to stop for an ARVN (Army of the Republic of Vietnam) checkpoint. They had pulled barbed wire across the road when they saw us coming. We stopped, and they told us the road was closed. We tried to explain that two trucks and two drivers were out there and that we *had* to get them. An argument broke out, and one of the ARVNs pointed his M-16 at the jeep. All of a sudden a dozen bullets hit him and his partner. We blew the whole place apart, and then went on our way. It was after dark when we found the two trucks. We checked them for booby traps, hooked them up, and took them away. When we passed the ARVN checkpoint again, it was just as we left it, but we kept going. A fool pointing his rifle the wrong way cost two lives, but that was the way things seemed to go. Stupid people (me included) did stupid things that cost human lives.

We were losing more trucks – and soldiers – to Charlie. Most of the time you never saw where the shells came from. Many of us were going crazy because we didn't know where to fire back. It's hard fighting an enemy you can't see. Sometimes they would shoot just one GI, and other times they would blow up a whole convoy. One day we came upon the site of a Vietcong ambush of an American convoy. The trucks were still burning, and the drivers had been crucified and were still hanging on crosses at the side of the road. I was scared all the time that I would die like that, and I began to hate every Vietnamese person I saw.

We were coming out of a village one day and an old man was riding a bike toward us. As he got close, I took off my helmet and bashed him on the head. He fell under my trailer and I ran him over. Every truck behind me ran him over, on purpose. Payback was ours – even though he had done nothing to us. We killed him (actually I killed him) to make up for those who had died. It was wrong of us to do this, but it felt good. He died in place of the gooks that we could not see.

Chi and I got married, by a Vietnamese village chief. There wasn't much of a ceremony. First he asked me some questions (mostly through Chi – I couldn't understand much of what he was saying). Then I gave him my dog tags so he could write my name and "New York/America" on the paper. Chi gave him her "papers" for proof of who she was. We went into a small stone building and he spoke and waved his hands, looking at both of us. There was a small Buddha in there with incense burning all over the place. When the priest smiled at us both and joined our hands I knew we were married. Just Chi and I were there – her family was in Saigon. Her mom was a prostitute, and she had a brother a year older than her, and a sister two years younger. She had no father.

My daughter was born in September 1970. I was on a convoy when she was born, so she was three days old when I first saw her. She was beautiful! Chi wanted to name her after me, but I took one "e" out. Le was her name. I loved her so much.

I began making arrangements to bring Chi and my baby back to the United States. Deep in my heart I knew that if I couldn't, Chi would become a prostitute. Women with mixed-blood children were the lowest of the low in the eyes of the Vietnamese people, because they want to keep their blood lines as pure as possible. So I don't think she would have had a chance doing anything besides selling herself. Her mom had sold her twelve-year-old sister to an American officer for $400 for one night. When I asked how she could do this to a young girl, her own daughter, so young, her response was something to the effect of: look what you are doing to my other young daughter, and I am not even getting any money from you! So I didn't like her mom at all. But I guess survival came first in Vietnam, no matter what.

I had to extend my tour of duty again. A year is a normal tour. I had already extended my tour by six months, and I wanted to extend it another six months. That way I would get out of the Army five months ahead of schedule. I went to see my commanding officer about extending my tour, and when I did, I told him I was married and had a new daughter. That is when all hell broke loose. He told me that he would not extend my tour because he thought I was going over the edge, mentally.

Also, he wanted to see my Vietnamese wedding and birth papers. I gave him our marriage papers, but I had no birth paper for Le because she wasn't born in a hospital. I was sent to see all kinds of other officers. Somehow, my marriage papers were "lost" and I was told I was not married anymore. I was told that the Army would not have recognized my marriage anyway, because they hadn't married us. I went through hell. I talked to more officers than I had ever seen before. In spite of all the bullshit, I did get my extension through. I guess someone liked my driving ability or something. So I had more time to work on bringing Chi and Le home.

Life was very strange now. I was full of hatred, but also full of love for Chi. I was going nuts. I didn't want anything to do with Americans anymore, but I had to keep trying to get the official papers to bring my family home. Whenever I went to a large army base I would talk to the highest ranking officers I could find. They gave me all kinds of excuses and stories but did nothing for me. I got more and more depressed. I even talked to the highest army chaplain on the base. He said he would look into my situation, but nothing happened.

Months passed quickly. Chi and Le couldn't move around much now, so I saw them less. And I was on the road even more. I guess the officers figured that if they kept me away I would cause fewer headaches for them. I was so angry that for the first time I wanted to kill American officers. I was hurting very badly. I didn't know what to do. My mind was in bad shape.

I tried to get work as a civilian in Vietnam. There were a number of companies there working for the American government. But each time I filled out an application, I was turned down: when they did the background check, my CO would tell them I was nuts.

About two weeks before I was to leave Nam, I was called into the CO's office. He told me to forget about bringing Chi and Le home. Apparently the chaplain I had talked to had contacted him and asked what could be done, but the Army had checked up on Chi and found that she had a shady past, that her brother was a suspected Vietcong. The boy was sixteen years old and hadn't joined the Vietnamese Army, so he was suspected of being a Vietcong. What a joke.

It was made very clear to me then that I had two choices: to leave them behind or desert the Army. I had been told all my life that the worst thing a soldier can do is to desert. Where would I go in Vietnam? As long as the Americans were there they would be looking for me. When they left, the Vietnamese government, communist or not, would get me. What was I to do? Get high. Nothing meant anything to me now – except Chi and Le.

Everything I had done in Vietnam was now catching up with me. Chi and I met one last time before I was supposed to leave. We both cried our eyes out. It was so bad, so much pain. We trembled in each other's arms. I left her and went back to my unit. Then she sent me a note saying to meet her at a cliff above the South China Sea, a very beautiful place where we had gone a lot. I went. I was leaving tomorrow, so I had to see her today. I took an officer's jeep and drove to the cliff. There they were, waiting, crying. We didn't talk, we just held each other, with Le in between us. We cried so much. I reached into my pocket and took out my pistol, put it to Chi's head and pulled the trigger. There was a splatter – then her blood gushed out – all over me. I held her tightly – with Le screaming still between us. I held her as long as I could – then let her go – over the cliff and into the sea they both fell. I pounded the earth as hard as I could – I screamed till I had no voice. I had nothing left inside me when I drove back. I should have died in Vietnam instead of living the thousands of deaths that I have.

Back at the hooch no one said a word to me. I had walked in covered with blood and looked pretty bad – no one said a word. I left Vietnam the next day. When the plane took off, everyone screamed with happiness – all but me. I didn't sleep, from the morning of the day I pulled the trigger till I got to Fort Lewis, Washington. We got to Fort Lewis at 2:00 a.m. I hadn't said a word to anyone for two days. I lay down and dozed off. Everyone got up at 6:00, went to chow, then started processing out of the Army. This took until early afternoon. I processed out of the Army without saying a word to anyone.

We were loaded onto a bus and taken to the Seattle-Tacoma airport where we boarded our planes to go home. After we took off for New York, a stewardess came to sit down with me. She started talking to me – I must have looked pretty bad – and she asked if I had just come from Vietnam. She wanted to know if there was anything she could do for me. I said, "Yes, please give me a needle and thread, so I can sew my stuff (ribbons and patches) on my army jacket." She took the coat and sewed everything on it for me. She treated me nice – the only one to treat me nice for two years. I fell asleep. And she woke me and gave me my jacket when we got to New York City.

I came home a very different person than when I left. My mind, heart, and soul were gone. I had sent some pot home from Nam in my baggage, so I stayed high all the time. I got a car, and within a few months married my high school girlfriend.

She was a high school senior when we snuck away and got married, on June 1, 1971. She had lived in five foster homes and wanted to get out on her own. I needed something, anything, to keep me from totally going over the edge. We married, but we didn't really love each other. It was more a marriage of convenience to us both. We didn't tell our parents right away, because we didn't want trouble for her with the child-welfare people who were responsible for her. Some weeks later she told me she was pregnant. She was working as a nurse, and I was unemployable.

Eight months later our daughter was born. Faith was born in April 1972. She would become my life for the next sixteen years. Everything I had left in me went to her. She was my little girl. The one to make right the wrongs I had done.

I hated myself. I didn't care about myself, I hurt so much. I couldn't stop thinking, dreaming, about what I had done. I died every time I shut my eyes. Chi and Le were there, covered with blood, inside my mind, in my thoughts. I started to have nightmares. I was a husband and father again, but how could this be – for the second time, a husband and father, again? I loved my daughter, but I still loved Chi so much that there was nothing for my wife. Can I ever forgive myself for what I have done? No.

I got a job just two weeks before Faith was born, at a water-filter plant. The town wanted a Vietnam vet to work at their plant, so I got the job. I worked a different shift each week, so my body never caught up. On the evening and night shifts there was five minutes' work each hour, and fifty-five minutes to think. I worked this job two years, and then I couldn't take it any more. I was numb in mind and body. Faith was my only reason for living. I did everything she asked me, except to stop smoking pot. I couldn't – it was all that kept me from death. I got high all the time. Pot was the only thing that kept my mind from burning out. That's the only thing she asked me to do that I couldn't. And I can't forgive myself.

My wife was in college most of our marriage. Working, studying, going to school, and sleeping was all she did. I was not very nice to her sometimes. I never hit her, but I did push her down, and I threw a plate of spaghetti at her once. I was mentally cruel to her, though. We argued whenever we were together. We would go to her friends' parties, and I would sit outside all night while she drank, talked, and danced. I hated to be around people, and she was a people-person. I guess I embarrassed her by not participating in her social events.

I didn't sleep much; I had bad thoughts on my mind. I was nasty, so I stayed away from people. There were a few I thought were friends, but in the end one of them even slept with my wife. I tried to trust her, but couldn't. As the years went on, I found out about some of the others she had been seeing. I really can't blame her, though. My wife was not a bad person; she had a rough life to live. She put up with more than most, living with me.

Once, during an argument, she hurt me very badly without even knowing it. She knew that Faith's middle name, Le, was the name of a child of mine in Nam, so she screamed out, "What's going to happen? Is some little gook kid going to come knocking at our door someday, hollering Daddy, Daddy?" That set me back years – I was without words. I just walked out, drove around for hours getting high. I had never told anyone about what I had done to Chi and Le, so she didn't know what I was going through. Things were never the same after that, and I went further and further downhill.

We had a son, Charles, seven years after Faith was born, and as my children grew, I had to spend more time around people. I didn't like that. Faith was a girl scout, so I had to take her to meetings and be around other parents. I didn't talk to them much. Then Charlie wanted to join the cub scouts. I took him to sign him up, and the scout leader said, "This is not the place to take a kid for baby-sitting; you have to actively participate with us." So Charlie never joined the cub scouts – I just could not participate with them. Faith did join a drum-and-bugle corps, and I would take her every week to practice at the local firehouse. I sat in my van waiting for her, three hours at a time, for two years – winter, spring, summer, fall. I could not be with the other parents waiting for their children.

As Faith grew, we were together all the time. Her mom went to college, worked, slept, and studied. Faith and I did the rest. I worked my jobs around

Faith's schedule and my wife's schedule. Faith and I did all the cooking, cleaning, shopping, laundry – everything. We were family – her mom wasn't. In seventeen years of marriage, my wife worked night shift a dozen or more years, so I raised our kids and worked at whatever I could. I could never hold a job for more than two years. I wasn't a very pleasant person. I worked by myself or with small companies with few employees so I wouldn't have to be around people. Twice I ran my own tractor-trailers. I didn't make much money, but it felt good.

This came to an end one day. As I was driving down a mountain, a tractor trailer in front of me slammed into the mountainside and caught fire. I stopped, and as I walked closer I smelled the smell of death. It took me back – back to the guy burning to death in my cab, the guy whose name I never asked. I drove my rig home, parked it, and I have never driven one since. But I still smell the smell.

When Faith was twelve, I bought her a horse. As soon as she learned to ride, she wanted to start showing her horse. She was a very good rider. I bought her a western-style show saddle for Christmas and she was ready. Early spring came, and we were off to the horse shows. Faith and her horse were probably the best thing that could have happened to me. I had to take her and the horse to all the shows, every weekend from April to November, so I *had* to be around other people. I would be there all day, and eventually people would come over and talk to me.

Soon Charlie started coming to the shows, riding lead line with Faith's horse. He had a little western outfit he wore. When I realized he liked his riding, I got him a pony. With both of them showing, I had to be around people. I met a lot of horse-show parents, horse traders, and tack-shop owners, nice people for the most part. I would talk to them, but I wouldn't befriend them.

Being at horse shows was good for me. I liked the horses and I think they liked me too. I would go down to the field and talk to them quite often. They listened to me, didn't pass judgment, and hung out with me just because they wanted to. When they were tired of me, they just walked away, without lying or making excuses. I would rather be with horses than people…

As the years went by, my problems got harder and harder to handle. My wife didn't like horses, so she didn't come to the shows. I think she was upset because we would be at a show every weekend and never did anything with her. I had bad dreams, couldn't sleep, and was working full-time as well as raising my family and doing all the indoor and outdoor household chores. I was tired of it all.

I couldn't trust anyone, so I never talked about my problems with anyone, and that cost me a lot. All my feelings built up inside of me, and finally something threw me over the edge. I fell hard, and in one second's time I lost everything. I was arrested and sent to jail. When I was released, I was ordered by the court to stay away from my family.

After a few days, I checked into the VA (Veterans Administration) mental hospital.[1] I was numb in mind and body. I was on suicide watch twenty-four hours a day. Something happened in my mind that just wiped out much of my memory. I was awaiting trial from August 1988 to July 1989. During those months, the state police picked me up twice more and took me to the VA hospital where I was put on the psychiatric intensive-care ward. This is a place where there is a line of rubber beds on both sides of the room, and a table and chairs in the middle. Nothing else. No books, no radio, no TV, nothing. Their intensive care was to leave you sitting, doing nothing, twenty-four hours a day. For one hour a day you would sit in front of a whole panel of doctors, psychiatrists, and other strange people, all asking you questions. Then you would go back and sit in the ward for another twenty-three hours. What a joke. You wore pajamas with no pockets, no shorts or T-shirt. You were watched twenty-four hours a day by the nurses, even when you had to go to the bathroom or shower. This was psychiatric intensive care.

Just before Christmas of '88 I got a letter saying I was now divorced. I had gotten a letter earlier saying that my wife had filed for divorce, that I should notify her attorney of my intentions. I had called her attorney and told her I would fight the divorce when I got money for an attorney. But a month and a half later, here were the final papers. I couldn't understand this. If it took two of us to get married, why didn't it take two of us to get divorced?

I was never invited to the hearing. So my wife got total custody of both children and all of our possessions, which included my birth certificate and social security card. Everything was gone. Never to see my children again. I called the judge who granted the divorce, and he told me there was nothing I could do about it.

There *was* nothing I could do. My mind was in a real bad way. And I had brought all this on myself by my actions in Vietnam. I got out of the hospital to go to jail for eight months, never to see my children again. This was July 1989. I was on twenty-four-hour suicide watch for eight months. I was given pills by the jail nurse to keep me down, and I had to talk to a counselor each week. But I had no will to live. The guards knew it and I knew it. Those eight months in jail were much longer than two years in Nam…

My problems are not the worst problems in the world, but they affect me as well as everyone around me. This is not just a story – it is a part of my life. And I don't know if I can ever forgive myself for my wrongs. I live from day to day and I am tired all the time – tired. Will it ever end? I don't see how. It's been with me over twenty-five years now.

2 Bloody Century

The trouble with much familiar talk about the lyric glory of war is that it comes from people who never saw any soldiers except the American troops, fresh, resilient, who had time to go over the parapet about once...Did you look, as I have looked, into the faces of young men who had been over the top...four, five, six times? Never talk to a man who has seen that about the lyric glory of war...Did Sir Walter Scott, Macaulay, or Tennyson ever see war? I should say not. That is where the glory of war comes from. We have heard very little about it from the real soldiers...

Harry Emerson Fosdick, *The Unknown Soldier,* 1933

World War I

Most history books tell us that World War I, the "Great War," began with the assassination of the heir to the Austrian throne by a Serb nationalist in Sarajevo. We are given the impression that this was a spark that ignited a tinderbox – as though war were inevitable, and the assassin, a tubercular young student by the name of Gavrilo Prinzip, was only playing to history's demands.

The war began triumphantly, ending a period of great faith and hope in human accomplishments and in the inherent goodness and integrity of the Western world. As Dalton Trumbo suggests, however, the reborn Age of Enlightenment seems to have had little lasting effect:

> World War I began like a summer festival – all billowing skirts and golden epaulets. Millions upon millions cheered from the sidewalks while plumed imperial highnesses, serenities, field marshals and other such fools paraded through the capital cities of Europe at the head of their shining legions.
>
> It was a season of generosity; a time for boasts, bands, poems, songs, innocent prayers. It was an August made palpitant and breathless by the prenuptial nights of young gentlemen-officers and the girls they left permanently behind them…
>
> Nine million corpses later, when the bands stopped and the serenities started running, the wail of bagpipes would never again sound quite the same. It was the last of the romantic wars…[1]

Woodrow Wilson was only one in a long series of U.S. presidents to deceive the American people and win reelection with promises about keeping America out of war. On August 19,

1914, Wilson cited the 1901 Declaration of London, stating that the United States would not take sides in the European dispute. He called on the American people "to be neutral in fact as well as in name...impartial in thought as well as in deed."[2]

The sinking of the British Cunard liner *Lusitania* by a German U-boat on May 7, 1915, began to turn the tide; among the 1200 who drowned were one hundred and twenty-eight American citizens. The torpedoing of the largest passenger ship in the world was portrayed as an act of unprovoked aggression, under which Wilson was "forced" to reconsider U.S. neutrality. In fact, the German government had published a formal warning to American citizens traveling on "belligerent passenger ships" a week before, indicating the *Lusitania* by name.[3] At the time, British and American leaders feigned horror that a passenger ship should be so targeted, but it is now known that the ship was heavily armed and carried thousands of cases of ammunition.[4]

The year 1916 was the most prosperous year in all previous American history. Following a severe recession in 1914, war orders from the European Allies had stimulated the economy and precipitated the sale of more than $2 billion in matériel. When Secretary of State Robert Lansing announced the likelihood of American involvement in the war, stock market prices soared to a 15-year high. Foreign trade soared to a record $8 billion, while domestic output peaked at $45 billion. America was fast becoming banker to the world, and its new status brought both power and arrogance. As historian Howard Zinn writes, "Governments flourished, patriotism bloomed, class struggle was stilled, and young men died in frightful numbers on the battlefields."[5]

Randolph Bourne later observed that "war is the health of the state." War was good business, but it was bloody, too, and the full extent of its bloodiness had to be kept from the public view if economies were to continue to thrive. In the first Battle of the Marne, 1,000,000 British, French, and German soldiers died. The unsuccessful German attempt to break through at Verdun resulted in the death of 600,000 men. Britain alone lost 20,000 men on the first *day* of the Battle of the Somme in 1916, as grim-faced generals flung ever greater numbers of human bodies over the trenches at a gain of mere yards. Altogether more than a million men, an entire generation, died in this one battle alone.[6]

That such carnage was kept well hidden back home is evidenced in remarks by British Prime Minister David Lloyd George himself, who attended a dinner party for the respected journalist Philip Gibbs late in 1917. Gibbs had just returned from the front, and his reports clearly shocked the British leader.

> Even an audience of hardened politicians and journalists was strongly affected. If people really knew, the war would be stopped tomorrow. But of course they don't know and can't know...The thing is horrible beyond human nature to bear, and I feel I can't go on with the bloody business: I would rather resign.[7]

Why the people of England couldn't know is left unsaid, and one can only wonder what they might have done if they had known.

In the United States, President Wilson began to quell anti-war sentiment as early as 1916, and already in his inaugural address he warned that "no faction of disloyal intrigue break the harmony or embarrass the spirit of our people."[8] "Rough Rider" Teddy Roosevelt used fewer words than the collegiate Wilson: "The man who believes in peace at any price...is not worth defending."[9]

In 1918, Congress added the Sedition Act to its Sabotage and Espionage Acts, mandating imprisonment for those who would "willfully cause or attempt to cause insubordination, disloyalty, mutiny, or refusal of duty in the military or naval forces of the United States, or...willfully obstruct the recruiting or enlistment service of the U.S."[10] An American Defense Society replete with Vigilante Patrols was formed to put an end to "seditious street oratory," and dissenters faced jail time for "criticizing the flag, government, draft, or arms production." Such laws were needed, it was said, to protect these institutions from "profane, violent, scurrilous, contemptuous, slurring, or abusive language."[11]

As Wilson announced a declaration of war against Germany in 1917, he waxed eloquent:

It is a fearful thing to lead this great, peaceful people into war, the most terrible of wars. But the right is more precious than the peace, and we shall fight for the things that we have always carried nearest our hearts...for democracy...for the rights and liberties of small nations, for a universal dominion of right by such a concert of free peoples as shall bring peace and safety to all nations and make the world itself at last free.[12]

British Foreign Affairs Secretary Sir Edward Grey was less enthusiastic, and envisioned not peace or safety but impending catastrophe: "The lamps are going out all over Europe; we shall not see them lit again in our lifetime."[13]

ARTHUR WOOLSTON, an 83-year-old Englishman, recently showed me a sheaf of yellowed letters sent home by his father from the front toward the end of the war. Arthur Sr. was called up in 1917 and trained at Blandford, in Dorset; from there he was sent on to France, where he fought in the trenches until he was killed by a shell on March 25, 1918. His wife was not forgotten by the British government, and at the end of the war she received a large bronze medal inscribed with the words, "He died to save civilization." But Arthur's letters, written in simple, local dialect, show no such grandiose delusions:

Dear Evelyn and the boy, July 1917

Again I send a few lines just to let you know how I am. Well I am putting up with it the best I can. My word you have to go through it with inoculation. I can't lift my arm up. We are supposed to have 48 hours off but we have been marching about after kit. This morning we went filling beds with straw ready for recruits. They keep coming as fast as they can…

Sept 10, 1917

Some started learning Louis guns this morning, others rifle grenades, so that makes me think that we shall get a fortnight at it. You say you don't think Artie had better have any cartridges, they are only empties they would do to play with. As for being a soldier, I shall do all in my power to keep him from a lot like this. I hope he will learn a trade that will keep him from it. A chap in our tent only said yesterday he will see none of his children comes in it…

My dear wife and Artie, Oct 20, 1917

We set off again yesterday for another train journey. We marched down to the station and it did weigh. I can tell you it's enough to break our backs. We had rounds of ammunition to carry as well. It's a job to lift it on our backs. We started from the station about dinner time and got to the station where we were to get out at seven o'clock. We had to march about eight miles or more to get to the place where we were putting up at. I don't know how we got there. It's the worst go I have had. I were beat by the time we got there. We never seemed to get any nearer. Everyone we asked kept saying the same so we didn't seem to get any nearer. We had to rest four or five times so you can tell we were glad to drop down anywhere…

We arrived here about twelve. They put us in barns. There are about forty or fifty in this one, we had no blanket or anything and we were starved. We are experiencing something now. We can hear the bombardment from here, some of our chaps got left behind and the rations too. We have had only two biscuits and bully beef all day today. I had four packets of cigarettes given since I have been out here. I give most of them away. They are twice as cheap out here…

You are all in my mind, after all our hardships I look on the other side of seeing you again shortly when this is over…

Dear Grandad and everybody and Evelyn and the *children,* Dec 5, 1917

I were very pleased to hear from you and to hear that Evelyn is over the stile now. I am pleased to hear both are doing well and I do hope they will, fancy a boy. I daresay Evelyn would have liked a little girl and I wouldn't have minded. I should never like to think little Artie and the youngster would ever become a soldier, they never will if I can rule…Let Evelyn have all she needs, never mind about money. She knows where there is plenty and she can have as much as she wants…I should have loved to have been at home and it was my place. It's a downright shame that we should have to be away. I shall look forward to the time of coming home more and more…

Jan 11, 1918

We have had a rough time since we have been here. I think it had been a bit quiet until we got here. My word we did get a shelling on the 30th. I shall never forget it. I shall have plenty to tell you when I come home. We lost about half the battalion. I think they were bent on making a big advance but they didn't succeed. I think we can get through anything after that, it's enough to wreck anyone's nerves.

We are having terrible weather for the trenches, my word it's freezing, fancy being out all night in this weather, don't know what to do with our feet. I am pleased we have come down for a day or two it's time we had a rest but we are still in reach of shell fire. We have such a job to walk after being so cold, just as we are getting warm it takes effect…

I am longing for the time when I shall be with you all again…I hope the leave I get will be to stay, and may it soon come. I shall do my part after this and I don't think I were bad before always at work but this will alter many a man. Some will be worse but I kept straight through it and mean to. I shall prosper in the end…It is about over shoe tops in mud and water today. God help them in the trenches…

Jan 12, 1918

We hear all sorts of rumors about peace and such like and when we shall be relieved but it's no use taking it in. We have to wait. Nothing certain in the army…

Jan 24, 1918

The first day we had in supports, well when we got there there were no where to get shelter and the trenches were terrible with mud...We walked right to the trenches on the top so you can see, fancy that they say they are Saxons in the front we were in and they don't want to fight. It were impossible to think about fighting under such conditions. We had to lunge on the side of the trench all night it did seem a long night. The next morning when it were light Fritz were walking about on top just as if there were no war on. He shouts out good morning boys, he were not more than twenty yards from us in his trench...

Later on several of Fritz's came and sat in our trench against us and started talking and giving us souvenirs. They looked very bad. They are fed up with it, they love to get hold of a piece of our bread. They don't like theirs much according to what they said. We said German bread good, they said no, and patted their stomachs. They gave me a cigarette and a mark in German which I am sending home for you to keep. Have a look and see what a mark is worth. The same morning I told you three of us went up the front line, well the other that were left in the place that we made got killed that morning. A piece of shrapnel went right through the zinc and hit him in the thigh. We were all very sorry. He were a good chap, a Welsh, and he had only just come from hospital...

March 11, 1918

I thought of you yesterday being Sunday. I fancied myself with you and the boys going for a walk as we used to. I thought what a shame it were to be throwing the time away in this manner...I hear Rumania has signed peace no wonder I think Russia has let her in I don't know what will become of it all. They will leave us in it to finish it yet. It's time someone looked into it and settled it. They haven't sent so many on leave the last three weeks but I hope they will start again so our turn will come quicker.

My dear wife and little Sunbeams, March 20, 1918

...We had fair weather lately but yesterday it broke it poured of rain and we were about up neck in mud again it's a terror it makes us tired out hardly lift our feet up. We had some experience at dodging minny werfers this time. I think they are the most deadly shells I have seen. We had warning by the Hoods that were there they had two blown to pieces the day before and that were the place we had to be, they happened to be inside a bivvy. You can see them coming in the air and we had to scuffle if they were coming toward us, the hole they made would bury a horse and cart.

I hope we shall get a rest after this time up so if we do you can stop parcel for a week or two, we may be able to get things if they have us back and I know you can do with it. I must thank you for the parcel it were all right and thank Fred and Elsie for what they sent and Julia it were all right, the fruit were very nice and the pudding extra. I have received a photo of our family and it is very good. I don't know if I shall be able to send this but I hope so. I remain with best love from your loving husband, Arthur.

Five days later, on March 25, 1918, Arthur was killed by an exploding shell.

PERHAPS THERE IS NO more fitting tribute to the desolation of World War I – indeed to all wars everywhere – than the closing chapters of *Johnny Got His Gun.* Published in the first week of September 1939, just days after the start of the next great conflagration, Dalton Trumbo's novel about soldier Joe Bonham proved eerily prophetic of carnage yet to come; take away the date, and many modern readers would set the scene in Vietnam. Minus four limbs and his face blown away by a shell, Joe is deaf, dumb, and blind, and must lift and drop his

head against a hospital pillow to tap out (in Morse code) his scathing message to all of mankind. He asks that the wretched trunk of his body be placed in a glass case and taken around the world on display:

Take me into the schoolhouses all the schoolhouses in the world. Suffer little children to come unto me isn't that right? They may scream at first and have nightmares at night but they'll get used to it because they've got to get used to it and it's best to start them young. Gather them around my case and say here little girl here little boy come and take a look at your daddy. Come and look at yourself. You'll be like that when you grow up to be great big strong men and women. You'll have a chance to die for your country. And you may not die you may come back like this. Not everybody dies little kiddies.

Closer please. . . Come on youngsters take a nice look and then we'll go into our nursery rhymes. New nursery rhymes for new times. Hickory dickory dock my daddy's nuts from shellshock. Humpty dumpty thought he was wise till gas came along and burned out his eyes. A dillar a dollar a ten o'clock scholar blow off his legs and then watch him holler. Rockabye baby in the treetop don't stop a bomb or you'll probably flop. Now I lay me down to sleep my bombproof cellar's good and deep but if I'm killed before I wake remember god it's for your sake amen.

Take me into the colleges and universities and academies and convents. Call the girls together all the healthy beautiful young girls. Point down to me and say here girls is your father. Here is that boy who was strong last night. Here is your little son your baby son the fruit of your love the hope of your future. Look down on him girls so you won't forget him. See that red gash there with mucus hanging to it? That was his face girls. Here girls touch it don't be afraid. Bend down and kiss it. You'll have to wipe your lips afterward because they will have a strange rotten stuff on them but that's all right because a lover is a lover and here is your lover.

Call all the young men together and say here is your brother here is your best friend here you are young men. This is a very interesting case young men because we know there is a mind buried down there...there you are young gentlemen breathing and thinking and dead like a frog under chloroform with its stomach laid open so that its heartbeat may be seen so quiet so helpless but yet alive. There is your future and your sweet wild dreams there is the thing your sweethearts loved and there is the thing your leaders urged it to be. Think well young gentlemen. Think sharply young gentlemen...

Take me wherever there are parliaments and diets and congresses and chambers of statesmen. I want to be there when they talk about honor and justice and making the world safe for democracy and fourteen points and the self determination of peoples...Put my glass case upon the speaker's desk and every time the gavel descends let me feel its vibration through my little jewel case. Then let them speak of trade policies and embargoes and new colonies and old grudges...Let them talk more munitions and airplanes and battleships and tanks and gases why of course we've got to have them we can't get along without them how in the world could we protect the peace if we didn't have them?...

Take me into your churches your great towering cathedrals that have to be rebuilt every fifty years because they are destroyed by war. Carry me in my glass box down the aisles where kings and priests and brides and children at their confirmation have gone so many times before to kiss a splinter of wood from a true cross on which was nailed the body of a man who was lucky enough to die. Set me high on your altars and call on god to look down upon his murderous little children his dearly beloved little children. Wave over me the incense I can't smell. Swill down the sacramental wine I can't taste. Drone out the prayers I can't hear. Go through the old holy gestures for which I have no legs and no arms. Chorus out the hallelujas I can't sing. Bring them out loud and strong for me your hallelujas all of them for me because I know the truth and you don't you fools. You fools you fools you fools...

He was the future he was a perfect picture of the future and they were afraid to let anyone see what the future was like. Already they were looking ahead they were figuring the future and somewhere in the future they saw war. To fight that war they would need men and if men saw the future they wouldn't fight. So they were masking the future they were keeping the future a soft quiet deadly secret. They knew that if all the little people all the little guys saw the future they would begin to ask questions and they would find answers and they would say to the guys who wanted them to fight they would say you lying thieving sons-of-bitches we won't fight we won't be dead we will live we are the world we are the future and we will not let you butcher us no matter what you say no matter what speeches you make no matter what slogans you write...

Remember this. Remember this well you people who plan for war. Remember this you patriots you fierce ones you spawners of hate you inventors of slogans. Remember this as you have never remembered anything else in your lives.[14]

World War II

FOR LEONARD DIETZ, World War II is not a happy memory. Now a retired nuclear physicist, he enlisted in the Army Air Corps in autumn 1943 and received his silver pilot's wings at age twenty-one. Mustangs were his favorite, and soon flying was as natural as breathing. After graduation, he was transferred to the 506th Fighter Group and shipped to Guam. A few weeks later, his group flew to Tinian.

His first view of Iwo Jima shocked him. The raw earth beneath his wings looked like a lunar landscape, pockmarked everywhere by shellfire. He could see no trace of anything green or living. On the ground, it was even more depressing: the stench of death was everywhere and could not be avoided. It persisted for many weeks and became especially bad after the earth was bulldozed or after a heavy rain, when body parts washed out of the hillsides. Twenty thousand Japanese defenders and six thousand eight hundred American Marines had died there, and twenty-two thousand more Marines were wounded in the ghastly battle.

Waiting in line on the tarmac before his first bombing run, the six 50-caliber machine guns on his wings armed and loaded, he realized he'd reached a turning point in his life. "Until this moment flying had been fun," he recalls. "Now it was time to pay the piper." In the air, his squadron intercepted two Japanese fighters. The Tojos dropped their external fuel tanks, rolled onto their backs, and began diving toward the ground. Then they split apart, going in opposite directions. Leonard and his element leader rolled into a very steep dive, closing on the Japanese pilots at well over 500 miles per hour. As he passed within range of one of the Tojos, Leonard fired a burst from his machine guns. A ball of flame billowed from under the fuselage of the Japanese fighter, and he watched as the pilot rolled back his canopy and tried to push his head and shoulders above the windshield into the slipstream. He was less than a hundred yards away. But the Tojo was in a shallow dive and began to spiral out of control. The pilot struggled

for two or three seconds against the tremendous force of the wind, but could not bail out. In Leonard's words:

Horrified, I watched him collapse back into his seat; moments later his airplane crashed into a hillside and exploded. I was breathing hard and my heart was pounding, even though I had my oxygen mask on. My eyes burned, and I was aware that my mouth was extremely dry and my tongue was sticking to its roof. I looked back at the thin column of black smoke marking the Japanese pilot's funeral pyre and felt remorse for shooting him down in such a cold-blooded fashion.

For Leonard and his fellow pilots, participation in the war became a bad dream that wouldn't go away. The atomic bombs were dropped on August 6 and August 9, but bombing missions over Japan continued until August 15, when Emperor Hirohito spoke to the Japanese people by radio for the first time, urging them to accept the Allied terms of surrender. Leonard recalled:

On August 10th, the day after the Nagasaki bombing, I took my last daylight mission to Tokyo to strike the Tokyo arsenal. There were three waves of B-29s, twenty bombers each. I remember looking down – we were at 23,000 feet – and we could see nothing of the city of Tokyo because it was leveled. It was like a giant hand had come out of the sky and mashed everything to the ground. It looked like it was hit by an atomic bomb.

The intensity of the air war in Japan was horrendous. The American public has not really ever been aware of the extent of the damage. It may have been reported in the press, but it didn't get much attention. More people were killed by fire bombs than were killed by the atomic bombs. Far more. This went on right until the 14th and 15th of August. The day of the 14th, there were a thousand airplanes over Japan, about two hundred fighters and eight hundred bombers. They struck targets all over Japan, whole islands, all through the morning of the 15th of August.

Years later, Leonard reflected at a Veterans Day service:

I learned that war is not glorious but is about suffering and dying. It is a vile and evil business that blights everything it touches, robs people of their lives and futures, and forces many good people to do bad things against their better natures. The whole thing was so bad, no matter how you look at it, that we can't ever let anything like this ever happen again.[15]

WHILE LEONARD DIETZ chased Tojos in the North Pacific, Siegfried Ellwanger was halfway around the world, fighting the Russians. Today he and his wife, Renate, are dedicated unconditionally to the cause of peace. All the same, he cannot forget his part in Hitler's *Reich;* he knows that history cannot be undone.

Born in Germany on August 7, 1924, Siegfried and his generation sought to avenge the wrongs of the past and embraced the new National Socialism with hope and fervor. Yet, as with generations of idealistic youth before them, their dreams were soon to be shattered. In an account of that time, he writes:

Germany in the early 1930s: A hopeless time of economic decline with all its consequences – unemployment, hunger, need and suffering in the families, alcoholism, crime, clashes of political opinions – Communists, National Socialists, and many other parties...

Smear campaigns against the so-called "enslavement of the German people" as a consequence of losing the First World War…And in the midst of this national chaos, a youth without a future, without any prospects.

Then, in 1933, Hitler's rise to power: Finally the "new time," promised for years, would become reality! Rousing battle songs of the Hitler Youth rang through the streets of the town, accompanied by fanfares, drums, and flags.

> With the flag of youth for freedom and bread…
> Our flag flies ahead of us
> Our flag leads us into eternity
> Our flag is more than death!

One year later I finally was old enough to join the march, like nearly all of my classmates. Who can imagine what went on in my ten-year-old heart, when for the first time in my life I experienced "community"? When I heard from all sides: you are called to great things! Together we will rule over the world! Together we are strong!

We wore our uniforms with pride. We could hardly wait to go to camp during vacation, and often on weekends, too. It was an experience of camaraderie and community, and it enthused us:

Long hikes, through the day and night, orientation with the help of a compass, or determining our direction by the sun…

In scouting games we fought "brown" against "red" and took "prisoners"…

Then, late into the night: campfires, setting up tents, singing, reading heroic stories. At six o'clock in the morning the blast of a trumpet, lining up in shorts, with our wash things, and jogging to the nearby stream. Then flag-raising, breakfast together, sitting in a circle, the beginning of a new day, full of expectation. Yes, we were the pride of the nation – and I was part of it!

And then: "The enemy is lurking at our borders, with nothing else in mind than to humiliate this proud nation…" Our holy fatherland, in danger! Who wouldn't want to be a hero, to defend this fatherland with all he had – even to stake his life for it?

End of August 1939: Headlines splashed across the newspapers and announcements on the radio: "Polish troops have crossed the German border!"

It was announced that on the first day of September we would all hear a most important speech by Adolf Hitler to the German nation. Everyone who lived in our house gathered in one of the big living rooms and heard Hitler shouting: "Since 4:45 this morning we have been shooting back!"

There was a disconcerted silence amongst the older people, and so we younger ones couldn't just shout with enthusiasm. I was fifteen years old and thought: "Why are they so silent?" Somebody whispered: "This is the end of Germany."

Early 1940: Within weeks Poland was defeated, Denmark and Norway occupied, France quickly overcome. Daily we heard of the victories, even on the eastern front in the fight against the Russians. Were Hitler's promises really coming true: a "German kingdom of honor, of power and justice – amen"?

When our youth leaders came home on leave – some of them already highly decorated, wearing medals for bravery – they told us, full of enthusiasm, about their experiences at the front. We younger ones began to envy them. When would it be our turn, my turn? Or will the war be over by then?

Sure enough, in 1942 my draft card arrived in the mail. I had just turned eighteen, old enough to be one of the fighters for the fatherland.

After a short training as a recruit in the south of France, our whole artillery company boarded a train in Marseilles, together with cannons, horses, and other soldiers. In the early morning of a cold day in December the train started rolling...

We passed through France, Germany, Poland, and on into the boundless plains of Russia. 1000 – 2000 – 3000 kilometers. After three long weeks – often interrupted by long waits – our train finally stopped beside the ruins of an old railway station. Darkness was falling. Suddenly we heard commands: "Get ready to unload!" Far off we heard the roar of the artillery. We had just reached the outskirts of the surrounded city of Stalingrad.

As we started marching to the front line, an ice-cold east wind hit our faces. The horses managed only with supreme effort to drag the guns through ice and knee-deep snow. The temperature was minus 35 degrees centigrade, and soon the first comrades had to be taken back because of frostbitten hands and feet.

Suddenly we passed a gully strewn with dead Russian soldiers – signs of a battle that had happened not long ago. Depressed, thoughtful, and speechless, we stood there for a short while. Why did these men have to die? Most of them were no older than we and would have had all of life before them. The first gnawing thoughts arose in me: what have they done to me? Why am I here at all – so many thousands of kilometers away from home?

In the midst of a vast field we set up our guns. No tree, no bush to be seen. We were far away from the next village. There were old abandoned houses where we sheltered our horses. We put up small tents for ourselves; a thin layer of straw covered the rock-hard, frozen ground.

Now and then we got a letter from home. They wrote about the "fight of the heroes in Stalingrad," where many hundreds of thousands of soldiers lost their lives. They also wrote about the merciless bombing of the German cities, where mostly women and children were killed.

Eventually we marched back to the Donets River, where we stayed for most of the summer. Our front was interrupted only by artillery fire. Then, at the beginning of autumn 1943, in the middle of one night – after we had fired all our remaining ammunition at the Russians, as a pretended warning of attack – we were ordered to retreat. The reason given was that the front had to be "shortened." In actual fact, the last big summer offensive of the Soviets had begun, which finally caused the breakdown of Germany's whole eastern front.

For me and for all the others, the high spirits quickly turned to deadly seriousness. During the nights we marched, hard-pressed by the Russian troops, often wading through mud and dirt. During the days both sides usually halted. And when night fell we went on – with our last ounce of strength. And still we hadn't seen the enemy. But that would soon change.

Autumn 1943: We lay alongside a road with our guns, hiding under bushes. Our infantry had already retreated. Our group of twenty soldiers had been ordered to stay until evening to fend off any Russian attacks. In the late afternoon we suddenly heard shooting from the distance. Then shouting, louder and louder: "Uraaah! Uraaah!" In a cornfield across from us we saw Russian soldiers in brown uniforms storming toward us. We only had a few grenades left, which we shot into the air to scare them off. Our situation was hopeless... Suddenly, about 100 meters away from us, they raised their arms and came running towards us. Was that a trick? No. They had no weapons and were surrendering. We took them prisoner and brought them back behind our lines. They were thirty-eight men!

But here let me tell what we experienced when we now saw them face to face: were these the so-called "gangsters," as it had been drilled into us? No! They were human beings like us, full of fear, sighing with relief and thankfulness that their lives were saved. And when we asked them to empty their rucksacks, we were flabbergasted: Chesterfield and Philip Morris cigarettes, cans with ham and eggs – American brand – and other things we never had seen before. Our officer proposed that we share these treasures with them...

On another occasion, we had gone into a small Russian village in firing position, shooting now and then to hold up the Russians, to irritate them. We stood around in small groups talking about letters from home, wondering when the war would ever end. With a good friend you could even sometimes talk about the senselessness of this war…

Suddenly we heard the sound of enemy artillery, hissing and roaring. Only seconds later we were in the center of a tremendous barrage. Within a short time sixty, eighty, a hundred grenades had landed among us. Then it was over, and from all sides came the terrible groans, the shouts, the stifled cries for help. The sight around us was terrible. The crying and groaning gradually stopped. Many were dead, many wounded. Why? For what? For whom? Some couldn't even be recognized anymore. And a few minutes ago we had been standing together, talking about the senselessness and hoping for a happy ending…

Summer 1944: Typhoid fever helped me out of all this. In my delirium I saw them storming on us again and again. There they lay, friend beside enemy. When I came to, I was lying in a hospital train, rolling back toward the West from Romania. We traveled all night, and there was no morning without a stop to carry out the dead.

Then we heard the news: the English and Americans have invaded France. They were advancing toward Germany. As soon as I could walk, I had to march again: "Continue fighting! The enemy is at the border."

In the meantime I had been promoted. Now I was in command of more than forty soldiers. In December 1944 we got orders to join the attack at the Belgian-French border. How many years had passed since we set off with such enthusiasm?

A wave of propaganda accompanied this final stand. Placards hung from the houses: "Victory or Siberia!" – while in the East, by early 1945, the Red Army was already on German soil.

April 1945: One day we had to defend a small village in the middle of Germany. Where were the people who had once lived in these houses? Where were the flowers that had once blossomed between these houses? Where were the fathers, the sons, the mothers with their children?

Many of these half-timbered houses were burning. We sought shelter in the basement of an old farmhouse. The shooting of the American artillery stopped, and then we heard machine guns and rifles. We peeped through a crack in the door. There they came running! "Americans," we whispered. The basement door flew open. "Hands up!" With rifles leveled, they led us toward a wood and into an abandoned quarry. "Shirts and shoes off! Faces to the wall! Arms up!"

The war was over, at least for us. We waited for the shots. But why did they examine us? Suddenly we understood: they were looking for that telltale, unmistakable sign. Each member of the SS had his blood type tattooed under the arm…

Finally we were allowed to turn around. We were still alive! In open trucks we were transported to France. On the way, we could scarcely recognize our country, which had once been so beautiful. Every town and village destroyed, bombed, scattered. Old women, sitting in the rubble, waving to us with tearful eyes. Everything broken, just like our hearts.

1948: After three years as a prisoner, I finally came back home to Germany. In 1946, the Americans had handed me over to the French. And now I heard the terrible reality: millions of Jews had been murdered, killed in concentration camps, tortured and tormented. Thousands of Germans had lost their lives, too – those few that had recognized the devilish regime long ago and had tried to rise up against it long before the war.

I looked back at the many events of the past fifteen years. Where had my youth gone? And what did my life mean, now that it had been saved? I was only twenty-four, a student with no occupation…

The Vietnam War

I'VE BEEN TO THE VIETNAM Veterans Memorial in Washington, DC, more than once, but after having worked on this book for over a year, I was eager to see it again. Perhaps "eager" isn't the right word – there is nothing pleasurable about contemplating 58,000 unnecessary deaths – but at the very least I would see The Wall with new eyes. I made the trip two days after Veterans Day. Brightly colored wreaths still lined the walkway along the monument, and a few letters and memorabilia were scattered at the base of the wall itself.

The lines of names go on and on, from Dale R. Buis in 1959 to Richard Vande Geer in 1975 – fifty-eight thousand names. Mostly they are all-American names, like Richard and John and Robert – but I noted a disproportionate number of Hispanic names, too, names destined to be on that Wall because they were born at the wrong time or in the wrong place.

I listened to people as they passed by, tightening their coats against the cold, soft-spoken and respectful, but for the most part seemingly unmoved. A fashionably dressed young couple seemed more intent on the wreaths of flowers than on the panels themselves, and an older man wondered, rather superficially, why the panels had omitted any mention of rank. An old woman sighed to her companion, "All these men. For what?" A child wondered innocently and persistently why "they" had to die, but was hushed by her mother, who snapped, "You're disturbing everybody."

Yet these questions must be answered if we are ever to close the door on one of the ugliest chapters in American history, and they remain largely unanswered after more than thirty years.

DON WAS CHANGED forever by Vietnam. I met him in August 1997 at the State Correctional Institution in Pennsylvania's Greene County, a sprawling complex laced with razor wire. Don is an inmate on death row. After signing in and emptying my pockets at the visitors' desk, I was directed to "go back" to the visiting area. I reached a set of heavy steel sliding doors and waited while an electronic eye acknowledged me and opened the door with a heavy clunk. The guard at the console, silent behind thick-paned protective glass, pointed to a dimly-lit cubicle at the back of the room. When I entered, Don was already standing there, separated from me by a heavy barrier of plexiglass. For an instant, he looked apprehensive, but then pretended to shake hands with me through the glass.

We couldn't swap stories for long without talking about Vietnam. Like so many others of his generation, Don had gone to the war thinking he was going to help people. But by the early 1970s, he was going to meetings of the Vietnam Veterans Against the War in Pittsburgh. Among the other veterans who attended, some were so traumatized they could hardly speak about what they'd seen and done. Don, however, wrote down a few of the experiences that haunted him, and his initiative sparked an entire collection of Vietnam journals, many of which are still preserved in a veterans' center in Monroeville.

He recalls one vet who never spoke at all, but who, after reading an account in one of the diaries, suddenly piped up, "Hey, that happened to me, too."

After my visit, Don copied and sent to me several pages from his collection of handwritten pages interspersed with neatly-pasted magazine and newspaper clippings:

On August 19th a young soldier arrived at the Bien Hoa Airbase in South Vietnam. Memories of Woodstock only a few days earlier tranquilized him as the big military transport rolled to a halt. The ramp opened, and the young soldier was greeted by a wave of oppressive heat and a sickening stench of rot and decay. In time the stench would soak into his skin, eat into his clothing, mingle with his food, and become normal – but never quite unnoticeable – nor forgettable.

Undaunted, he hoisted his pack and seabag onto his shoulder and walked down the ramp. He was destined for his assigned fate somewhere in the jungles and he was eager to face that challenge, to follow his illusions. Ready to serve his beliefs, beliefs for which his small-town patriotism was as much to blame as his youth. With his pack and seabag he carried a conviction that he was an honorable soldier, that his country needed him, that he was doing something noble and good. He kept the pack and seabag throughout his tour in Southeast Asia. The conviction, he lost – left behind with his youth, his ambition, and his soul.

Third day in-country I reached my assignment at Can Tho in the Mekong Delta. The Can Tho army airfield – bigger than a bread box, smaller than a city block. An oval-shaped perimeter of barbed wire, water-filled trenches and craters, claymore mines, and sandbag bunkers – surrounded by thick jungle swamp and wide, muddy rivers. A real vacation spot. Narrow PSP walkways connected the shacks, hooches, and sandbag bunkers – metal sidewalks. I stepped off and immediately sank up to my knees in mud. My new assignment was with the "Loach" Scout Team Air Crew of the 271st Assault Helicopter Group, which was attached to the Search-and-Destroy Battalion.

The 271st consisted of about 130 men and 16 armed helicopters – Loaches, Cobras, Huey Gunships, Huey Slicks, Chinooks, and a Kiowa. It had 14 combat air crews, each assigned to its own helicopter. Its headquarters and support staff stayed at Can Tho. The air crews and helicopters stayed in the field, living at different camps around the Delta, moving where they were needed, returning to Can Tho only for periodic maintenance inspections. Forty-five men with no base – no permanent home – working in bunches of threes or fours here and there, eating C-rations, living in their clothes, sleeping wherever they could. A pretty tight-knit bunch – by necessity.

I had never been stoned before in my life. Didn't even know what marijuana smelled like. I had finished my three weeks of orientation and scout training. They taught me how to stay alive in a Loach: always fly low and fast, never hover. They gave me my own Loach and crew and sent us to Fire Base Grant with two other Loach teams. We had been there a couple of days. Hadn't seen any action – just spotting for artillery.

One night I was in the team bunker, and my gunner said he wanted to bring a couple other gunners over and get stoned – did I mind? I said why not – I've been to California.

You hear about people who don't get high the first time they smoke. Whoa! Not true. They passed a pipe around, and I woke up the next morning with my feet down in the bunker and just laid out straight back. They had taken my brain out the night before and passed it around and played with it – they had fun with me. They were my flight team and it was the first time we got to know each other. It made a big difference, becoming part of the team.

They certainly initiated me.

The Siege of Cao Lam

We were at Cao Lam, a mud-hole outpost on the border. The camp was triangle-shaped, about the size of a football field, maybe slightly bigger – a wide mud wall for a perimeter – fighting holes dug along this wall and mud bunkers at the three corners. It was home to an ARVN company with American advisors. We were there to scout the border – two Loaches and a Slick – ten men.

I was lying on the wall in about an inch of water, bone tired – half sleeping, half watching the mist float on the river. We'd heard reports of enemy movement: NVA regulars just across the border in Cambodia – a few hundred well-trained, battle-tested enemy. We had lost Varnado and his crew that day – in a ball of fire on the side of a mountain. Lying there, a song Varnie always sang kept running over and over in my mind, a song about aviators in World War II – we live in the air and die by fire…da da da…

Suddenly there was a loud, cracking sound, like lightning, and the wall shook. The wall was really just a built-up mud embankment, and I sort of instinctively rolled over into the hole I'd picked out for myself. I don't know why – I didn't know what was going on but I had a feeling it wasn't good.

A few seconds later there was a whistling sound, like a high-pitched scream, and the wall shuddered again – harder and more violently. Then mud and debris started falling on me, and I heard someone yell: INNNCOMMMING!

And then it began – a furious concentration – the fiendish wail and screeching of short artillery and mortar rounds dropping into the camp. A diabolical howling of torment. An enormous and continuous roar that went on for a full five minutes – it was a lifetime. With each explosive impact the ground bounced me – actually slammed against me and bounced me in the air. I could hear hot shrapnel zinging – pinging – wooshing – over my head. I flattened myself out in the hole, pushing myself deeper into the mud, trying to become a part of it. Movies on TV hadn't prepared me for the reality. I grabbed handfuls of mud in a feeble and pathetic attempt to hold onto the earth – a fool's anchor. I was afraid of being bounced out of my hole up into the murderous space above.

A few moments of silence, and then again the savage wailing salvo – tormented demons – and the intense roar. Blast stung my side – heat singed my hair – the ground slammed violently up against me and I bounced at least six inches straight up. I heard someone let out a cry – the pitiful cry of a frightened child – and then I realized I was alone in the hole.

In the first minutes of the siege our helicopters went up in flames, bright orange balls that lit up the night. Sappers had snuck in and tossed satchel charges. We never saw them.

The siege continued for eleven days.

Helicopter Assault

How does one explain a helicopter assault – flying straight into an enemy position at 100 mph, 10 feet off the ground – all your guns blazing at him, all his guns blazing at you? There's a lot of excitement in it the first time, a strange excitement. After that it becomes extremely unpleasant. When engaging the enemy on the ground you have some control over your destiny, or at least an illusion of it. But in a helicopter you don't even have that.

The noise, the speed, the enclosed space, the sense of helplessness – total helplessness. These emotional pressures are so intense you clamp your teeth until they hurt. You are pulled and shoved in so many different directions all at once by an incredible range of extreme, conflicting sensations – the indifferent forces of gravity, the jarring twists and thuds of ballistics, the screaming machinery. In the cramped cockpit claustrophobia claws at you. You're trapped and powerless, and you can't breathe. It's absolutely unbearable. Yet you have to bear it. Pull out too soon, and you expose your underside to the enemy guns, before you can kill him – and you die instead.

Your mind screams to be through the run and out the other side – to do it quickly – but that desire is twisted by the danger that's waiting there for you. Still you are attracted by the danger – drawn to it – because you know it's the only way out. That's all you want, so you hold the approach – coming closer and closer. You see the enemy guns flashing – licks of red-orange flames directly in front of you, red and green tracers arching toward you – hunting for you. You slump in the seat, instinctively pulling in your elbows and knees, hoping it will make you a smaller target. And the screaming increases. You know you are now screaming out loud, at the top of your lungs. You don't hear it with your ears. You just feel it. It's the fear inside you screaming to escape – to flee. A fear so deep and so basic it makes you cold and you shiver in the 100 degree heat. An absolute terror so overwhelming that your body separates from your mind. It takes over, reacting instinctively. Your stomach tightens – bile burns your throat – your mouth is too dry to swallow – your vision tunnels – you lose control of your bladder. And with it comes an uncontrollable rage deep inside you, a blind fury toward that force which makes you powerless: the enemy. You can see him now. You can see the fear on his face. You recognize the look. It's the same fear. He knows what you know – in a matter of a few seconds one of you will be dead. There's no other choice. No other way. Later, if you live, he'll revisit you – a slow motion movie in your mind. He may have been a good person, maybe someone's father or brother. He may have been a beloved teacher or gentleman farmer. In another place and time you might even have been friends. But at the moment you have to take his life – or he will take yours. Even if he doesn't want to. And when he visits later you will feel a heavy sadness, wondering who really is the loser – him or you.

The rage builds inside you, becoming a beast unleashed. And it focuses on the enemy, on his fear, and it becomes a fierce resolve to fight – to kick, gouge, hit, claw, bite – until the danger is gone. Until you have killed a human you did not hate, a gentleman farmer who could have been your friend. They call that courage and award it with medals, but it can never be separated from the fear that aroused it, or from the madness that accompanies it.

All of these emotions are so intense and so painful you can't handle it very long. All you can think about is ending your anguish. You don't care about rights and wrongs. You don't consider life and death. You give no thought to the chances of victory or defeat, or to the battle's purpose or lack of purpose. All that matters is the final critical instant when you release yourself into that violent catharsis you both seek and dread – and fly out the other side.

Now jam all this into twelve seconds – the average time of a typical helicopter gun-run.

Windows into Vietnam

I saw my first dead person, my first dead human being. I watched him die – saw the life go out of his eyes. He was running past my hole. A mortar round went off and he dropped right there, right in front of me, no more than two feet away. He looked at me and asked if I would help him. It was all very calm. No sense of urgency. I said sure, I'd help him – he'd be fine. And then he just died – very calmly, very quietly. And I guess I felt something inside me die with him.

His name was Billy Boy. He was nineteen years old. From the waist down he had no body left. Barely enough for his family to bury.

Water was our biggest problem – there wasn't any. A camp enclosed by a slimy mud embankment, pools of stagnant run-off everywhere you crawled, surrounded by swamp and rivers – and we had no water. Pure irony. We took water from the pools, boiled it on hobo fires, and dumped in lots of C-ration instant coffee to mask the taste and smell – to conceal anything in the water. During the brief lulls we brushed our teeth and washed our feet with cold swamp-water coffee.

They tried to resupply us. Big C-130s would roar over and drop parachuted packages. Two guys were killed – crushed to death in their hole when a supply pallet dropped on them.

Once, I was sent to MAC-V Headquarters to attend an awards ceremony put on for the press. I was to receive the Vietnam Cross of Gallantry.

As I stood there in my dusty nomex, mud still in my ears, and in desperate need of sleep, I listened to the brigade commander's words of glory and praise and how war is hell but necessary. And I began to hate that fat little shit standing in front of me in his clean, fresh uniform and his neat haircut. I hated him because he sent young boys into steamy jungles to die while he relaxed in the comfort of his plush, air-conditioned office. Who the hell was he to speak of brave men and their total dedication fighting for a free Vietnam!

So, I pretty much expressed my feelings by pissing on the back of his leg. He turned and looked at me in shock – not anger, just shock – and there I was, eyes straight ahead, at a perfect stance of attention, pissing over his feet. The press people were having a field day. When I finished, I gave him a smart military salute, turned, and marched away. No one said STOP. No one said a word. The only sound heard over the clicking of cameras was a cheer from a company of grunts who had been made to stand the formation. And I just marched away.

I walked to the flight-line, climbed into the first helicopter and simply flew away – no crew, no clearance, no flight gear, no Gallantry Cross.

When I got back to Can Tho, I started to cry – and I couldn't stop. For two days I wept continuously, never stopping. All I could think about were the people I had killed, about the good men I saw die, about my buddy – and I kept crying until our company doctor sedated me. I was taken to 3rd Surg where I told a doctor everything I had seen and done in the past nineteen months. All he did was put me on 60 mg of valium a day and lock me in a room for two weeks…

One day Chau Doc was hit hard. All hell broke loose on the perimeter. Tracers flying everywhere, guns roaring and popping all along the line. Everybody yelling and screaming. A couple of VC sappers were shot trying to blow the wire – their lifeless bodies hung over the posts. And it was raining the whole time, coming down in solid sheets.

Then suddenly the rain stopped, and there in the sky was this beautiful rainbow – the most beautiful thing I'd ever seen. And at that moment I began to hate what we were doing in Vietnam – began to hate myself for believing in it, for being a part of it.

I heard someone in the bunker say out loud, "Hey God, you trying to tell us sumpthin?"

People back home had forgotten us. They considered us murderers, animals, misfits. They didn't want anything to do with us, just left us to live or die – whatever – so long as they weren't involved. They didn't think about the fact that they had sent us there. We were to blame – not them.

But in Vietnam we helped each other, cared about each other. And we respected each other. We were all we had. We didn't fight for any ideology or higher purpose. We fought simply to survive – nothing more. And we looked out for each other. Men who fought together in Vietnam loved each other, without ever saying it. We were brothers in the truest sense.

My friend died in Vietnam. He was murdered. Pieces of hot jagged metal ripped his body to shreds – and I don't know who to blame. I escorted him home for burial. A big gray casket. I sat with it the whole way, touching it every now and then to see if it was real – hoping it wasn't. I worried about what I would say to his family, hoping it would be positive, something they would remember and pass on. But they wouldn't talk to me – didn't want to see me. They sent a minister to tell me not to attend the funeral. They didn't want me there. I didn't blame them.

I wish I could remember what it was *you* said…It wasn't important or anything, but it haunts me. And if I could remember, I'd write it down on this page, and maybe make you seem as alive to others as you still seem to me.

One day I was just sort of hanging around the command bunker at Cao Lam, watching a group of new recruits. They had just arrived in Vietnam. They were all joking, smoking, playing grab-ass – it was pretty loose.

All of a sudden, four helicopters came swooping in – not even touching the ground. They just started dumping bags from about five feet off the ground. One of the bags broke open when it hit the ground, and what came out was hardly recognizable as a human being.

The new guys stopped laughing. Nobody was saying anything. Reality had suddenly gone beyond words. A couple of guys were shaking, some started throwing up, and one guy got down and started to pray.

And I said to myself, "Welcome to Vietnam, guys."

Newsweek Jan. 1970:

> Sealed Off. The destruction of Ben Suc, a Saigon River village complex that supported the Viet Cong, was typical. It took only a minute and a half for 60 helicopters to descend on the village with a battalion of the U.S. 1st Division. While loudspeakers warned residents to stay in their homes, infantrymen quickly sealed off the town, catching many of its Viet Cong defenders by surprise. The villagers were assembled and the men between 15 and 45 led off for questioning. Within three days, Ben Suc was deserted, its people and their possessions loaded aboard boats and shipped twelve miles downriver to a refugee camp until they can be permanently relocated. Shortly after they left, torches were put to their homes. After Operation Cedar Falls ends, it will be a long time before the Viet Cong, or anyone else, will be able to use the Iron Triangle again.

I remember looking down between my feet and seeing the embers of desolation not more than 20 feet below the helicopter. Smoldering craters, human bodies, and burning skeletons of hooches – of homes. And I remember a guilt pressing down on me so heavily I could hardly breathe. I was sick of it all – sick of war, of what it was doing to us, but mostly I was sick of myself.

It wasn't just the senseless obliteration of Ben Suc Village that bothered me. It was the dark, destructive emotions I felt during the assault, from the moment we slipped in over the treetops through the morning mist lifting off the sleeping village – urges to kill and destroy that seemed to rise furiously from the fear of being destroyed myself.

I had enjoyed killing the three Vietcong who ran from the tree line near the village. Feeling like a glorious bird of prey swooping down, I watched the mini-gun rounds splash through the paddy toward the running men, then ripping and tearing their bodies to lifelessness.

I remember the strange sensation of seeing myself as in a movie – one part of me doing something while the other part watched from a distance, shocked by what it saw, powerless to stop it.

I was awarded an air medal for my actions that day.

For heroism, they said.

How absurd, I remember thinking.

I can analyze myself all I want, relive the horrors of Ben Suc Village for a million years, as I feel destined to do, but that won't make a new village rise from those ashes. It won't answer the questions that still burn in my mind. Nor will it lighten the burden of my guilt.

Doi sau…

Two American jets had just napalmed a grazing cow and the three of us in the Loach laughed. Then the jets zoomed in again, dropping more napalm. They made one low pass and we cheered. An hour later we landed for a pickup, and I got to see the survivors. They were carrying their belongings and they had no place to go. It was like a parade across the rice paddy – a string of 75 or 100 peasants carrying burned and screaming babies, women, old men, some badly cut up with shrapnel wounds. When I got back to camp I got drunk for the first time in my life.

Word quietly comes down that a Christmas cease-fire has gone into effect. From the darkness of the perimeter you hear "Ain't Christmas fun out here in the mud and the rain and the shit!"

There's a loud roar and a hard, hot slap of air, great pressure, and something slams you viciously in the back. You go facedown in the mud, ears ringing, a familiar sensation of having your mouth stuffed with cotton. You hear a carbine rattle off a few automatic bursts, and someone screams: "Medic, Medic!" It all sounds so far away.

A few yards behind you is a scorched, smoking crater of earth. Lying beside it is a soldier. He slowly gets up on his feet and then falls when one leg collapses under him. He lies there flopping one arm back and forth saying, "Boom, boom…"

Beside you is another soldier, still down in the mud, mumbling. There's blood oozing from holes in the back of his head, and you numbly try to hold a compress over them. Your hands turn red with his blood.

Lying closer to the crater is another wounded soldier – his face peppered with shrapnel. A mass of raspberry red. His partner fans him and cries, "He keeps goin' out, man! Medic!! He's gonna go out for good, man!"

And then you see your own partner lying on the other side of the crater – the Doc over him. And you realize with gut-twisting horror that he has been hit. In your blind run to his side you notice gauze and compresses covering his chest and stomach. The compress over a hole torn in his chest is soaked with blood. Every time he breathes, pink bubbles form around the hole and burst. He makes a wheezing sound and tries to talk but can't – his windpipe fills with blood. But it's his eyes that frighten you most – the hurt, dumb eyes of a child who has been severely beaten and doesn't know why.

You feel the separation in his eyes and you know he's alone in another world – isolated by a pain you can't share with him and by a terror of the darkness that's swallowing him. His eyes and his silence, and the foamy blood and the wheezing gurgle in his chest, arouse in you a sorrow so deep and a rage so powerful you can't distinguish between the two.

You sit there helpless, hurting, the rain falling on you all the time. You watch your friend fall into unconsciousness. And you almost envy him…

Merry Christmas, Brother.

On the night of April 26, 1970, five helicopters of the 271st airlifted a company of the 44th ARVN Rangers into the village of Prasant in the Parrot's Beak of Cambodia. I flew gun cover for the assault. The real killings of the Cambodia invasion began that night – four days before the actual invasion – but the government never told of those deaths.

The Rangers swarmed into the village and simply shot everyone. Afterwards, at daybreak, they threw the bodies into the river. Then we assaulted the village of Changwar. Again the villagers were all shot, their hands tied behind their backs, and the bodies were thrown into the Mekong.

We followed this pattern of destroying villages for the next three days, and by the time the invasion was initiated on April 30, 1970, the Mekong River was clogged with swollen bodies floating downstream. The water was the color of rust.

Sooner or later it got to be too much for all of us. I saw men deliberately try to get wounded. I saw one man casually stand up during a fire fight and get it through the head. Another stood up too, and only got it in the arm and leg. He was sent to a hospital in Japan, and everyone thought he was the luckiest guy in the world because he was safe and free again.

I remember one gunner – a young PFC from Michigan. We were going out to fly an assault mission and this gunner climbed into the Loach, strung his ammo belts, braced himself with his M-60, and never said a word. But he was trembling – from start to finish of the mission, all day long he trembled. I kept quiet. We didn't laugh at him. We didn't dare. It was "there, but for the grace of God, go I."

It was his last combat mission before going home, and he was in absolute terror.

I actually came to enjoy combat. It seemed only heavy combat and killing could make me feel so alive. I would fly into situations others thought were insane – certain deathtraps. And slowly I came to understand that I didn't enjoy it, I'd only convinced myself that I did – to justify killing other humans. It was a way of denying the guilt I felt – that I hated myself for what I was doing. I eagerly went into those death-trap situations because I was trying to kill myself – to end the anguish. Others saw it and refused to fly with me.

They took me off the Loach Team roster – said it was almost time to go home, said I'd be safer hauling troops and supplies – no more gun-runs. So they put me on the Chinook Team roster.

I had certainly been down the road. I had fought in some nasty battles – U Minh, Iron Triangle, Parrot's Beak, Cambodia, Cao Lam... I had flown gun for the 1st Infantry, 9th Infantry, 25th Infantry, 5th and 7th ARVNs, 44th ARVN Rangers, and some unidentified cloak-and-dagger outfit of some kind – black ops, they called it. I had flown combat assault, gun cover, hunter-killer, scout, recon, river patrol, and once I even flew armed escort for the Miss America show which toured the Delta. I slept in all kinds of places – Can Tho, Dong Tam, My Tho, Chau Doc, Cao Lam, Di-An, the Plantation, Fire Base Grant, Tay Ninh, the Iron Triangle, Outpost Spider, Black Virgin Mountains, Cambodia, and a few places that didn't even have names. Over 2000 combat flight hours. I felt like an old, old man. I was twenty.

Finally I fell apart – complete emotional collapse – and they sent me home. Not because I fell apart, but because I had nothing left. I was of no use to them anymore. They had sucked everything out of me – all I had – my strength, my humanity, my youth – I had nothing left, so they sent me home. Okay, we're done with you – you can go now. So they sedated me and took me to 3rd Field Surg in Saigon. A doctor wrote in the record: suffering physical and emotional exhaustion, malaria, and drug addiction; subject weighs a mere 118 pounds; recommend release from combat status. So they sent me home. They didn't fix me or help me or anything – they just pumped me full of valium and dumped me in the middle of Main Street, USA.

I thought, "Wow. Unbelievable. I'm home, really back home." But then people started yelling at us. Someone spit at me. They were calling us murderers, trying to pin the whole thing on us as if we were personally responsible for the war. I just couldn't take that kind of a reception, and so I hopped right back onto the next plane for Nam. I knew there were guys who understood me back there. We understood one another…

TALKING TO COMBAT veterans is a little like viewing the same landscape through different windows of a house: certain elements remain constant, but tints and shadows change and perspective alters. Fred Louis opens his window into Vietnam:

I was less than a month in-country, very green. We were on patrol in a free-fire zone. As we walked along a ridge, we spotted a few people in black pajamas and coolie hats walking on a trail in the valley below us. The whole company, over a hundred men, lined up and started shooting. My weapon was an M-79 grenade launcher, so they were well beyond my range. I stood and watched, stunned, bewildered, afraid, and fascinated by this somewhat bizarre shooting gallery. A buddy offered to let me shoot his rifle. I said, "No, that's okay." A small patrol went down and found a young boy with a belly wound. They left him. Our captain called in three VC kills to headquarters. A bitter taste of my year to come.

Later on I watched my squad leader check the papers of a wrinkled Vietnamese man walking with his granddaughter. His head was bowed in fear, eyes darting. My sergeant handed the papers back and said, "Go." We watched him walk away in relief and apprehension. Then my squad leader raised his rifle and shot the man in the head. You must understand something about M-16 rifles: they are small caliber, just slightly bigger than a .22, but the muzzle velocity is extremely high and the barrel rifling is specially designed. This causes the bullet to strike a human body with enormous impact and to tumble around once inside. So the man's head exploded like a watermelon hit with a sledge hammer. Small chunks of brain, white with flecks of blood, landed on the toe of my boot. I remember gently brushing away those pieces with the back of my fingers. That's all I remember. I shut down some more. Must survive.

I have taken over twenty-eight years to cry for the old man, for his granddaughter, for my sergeant, for me. I did nothing. That's what you did. Nothing. SOP, Standard Operating Procedure. You depended on your comrades for your very life. They carried loaded weapons and much anger. Little My Lais – little massacres – happened hundreds of times a day in Vietnam. You must remember that it does not take monsters to commit monstrous acts. My squad leader returned home a decorated hero, a Bronze Star and a Purple Heart on his chest. You may know him. He may deliver your mail or wait on you in the hardware store or teach your children. Or he may be stalking the jungles of Hawaii or some other forest with other men unable to return home. Or he may be one of the 100,000 suicides since the end of the war.

I must apologize. I gave you wrong facts. I don't remember if my sergeant used a rifle or a pistol. I don't know if the old man and the young girl were related. I brushed someone's brains off my boot, although I don't know if it was that day. But what I have said is still absolutely true.

Fred passed on a piece by Bruce Cole, a friend of his who had been a Marine Company Commander in '66 or '67:

One of his hands had just fallen right off, burned off. Now his charred wrist bone was sticking out from under the poncho we'd wrapped him in to drag him here. We had to use a second poncho to cover where the first one had melted, he was burned so hot.

PFC Wall died today. Burned to death in the service of his country. A blond, fair-haired kid whom everyone loved. Didn't even shave yet. The other men would tease him: "Hey, Wall, let me borrow your razor, ha-ha-ha." Now, sitting here, everything is very still. Sunset, orange sky, red clouds moving slowly. Water buffalo knee-deep in the pink-orange water, grazing on tender green marsh grass along the shore. Fishermen's sampans anchored out in the lagoon. Fish cooking, smoke rising, no wind. Everything calm, so serene. How do they go on living day to day like this, with the war going on all around?

Now I can hear birds singing all around. Dark-green mountain forest reflected in the mirror lagoon. PFC Wall's cooked body is cooling off. The sick-sweet smell is going away. Now it's too late to fly his body out. His charred, contorted body is lying here next to me. He's a "routine medevac." That means we don't send the dead out before the wounded. The wounded have higher priority. It makes grim sense in this insanity. Too many wounded today. His body stays with me tonight, burned black, cooked meat.

Now the fishermen's fires seem brighter. The sun is gone, the birds have stopped singing. Darkness comes. Swoosh! Crack! Incoming! Fuck! Rockets from uphill. Blinding flashes, shrapnel cracking, men screaming. "Doc, doc, where the fuck is doc?" Doc's hit. The first rocket hit his position. Crack! Crack! Now I hear mortar tubes thumping out there. Rockets and mortars.

Can't do a fucking thing but lie here, face down, caught in the open. Hear shrapnel cracking into trees, smell the broken trees. Each explosion comes closer. Fuck. They're walking the mortars across our position. Each one closer, each one louder. The ground jumps higher, dirt and rocks fall on my head. Hot shrapnel. No helmet. Shit. The next one's coming up my ass. Jesus Christ, I'll kill those fuckers. Just let the next round pass me. Come on! Come on! Come on! Crack! They miss me. I breathe the delicious smoke and dust.

I am lying with the top of my head in PFC Wall's armpit. The meat is falling off the bones. It's stuck in my hair. I wonder if his body took any shrapnel from me. I don't want to know. I cover him up. A man is crying for his mother. Gurgled coughing. Muffled sobbing in the quiet of the night. Too quiet. Assault coming soon. I can feel it. I know it. Too quiet. Got to get our shit together. Perimeter defense. Redistribute ammunition. I will never again remember the cooked flesh stuck in my hair. I never want to. I never will.

I remember now, twenty-six years later, that…that…I shaved my head.

The Falklands War

SINCE 1833, THE WIND-SWEPT ridges, peat bogs, and rocky outcroppings of the clustered Falklands Islands have been in British hands, despite Argentina's repeated claims to them. The two major islands of the Falkland archipelago, East Falkland and West Falkland, lie in the south Atlantic, some 300 miles off the Argentina coast. Together with about 200 smaller islands, they form a total land area measuring approximately 4700 square miles. Lying from 700 to 2000 miles east and southeast of the Falklands are the South Georgia and South Sandwich islands. The combined islands' population was estimated at 2100 in 1991.

In 1982, faced with Argentina's worst economic depression since the 1930s and mounting protests against his military junta, General Leopoldo Galtieri ordered the invasion of the Malvinas (the Argentinian name for the Falkland Islands). Under the command of General Mario Menedez, the first forces of *Operación Rosario* landed on April 2, quickly overpowering the island's small detachment of Royal Marines. A day later, Argentinian troops seized the island groups of South Georgia and South Sandwich. As he had hoped, Galtieri watched the pendulum of popular opinion swing convincingly back into his camp.

Diplomatic efforts to avert the crisis achieved nothing; everyone seemed to be spoiling for a fight. British Prime Minister Margaret Thatcher responded swiftly, ordering a naval task force to assemble and set sail. From the mid-Atlantic British-held Ascension islands, the Royal Air Force launched a 1000-mile bombing raid – one of the longest ever – on the airfield at Port Stanley, the Falklands' capital.

Pausing at the Ascension islands only long enough to regroup, a task force of fighting men and their machinery closed on the Falklands, harried by the Argentinian airforce. On May 21, the first British troops made an amphibious landing on the northern coast of East Falkland, near Port San Carlos, and the ground war began.

In all, the war would claim the lives of 1001 men – 255 British, 746 Argentinian – and leave nearly 2000 wounded.[16] It ended on June 14 with General Menendez signing the terms of surrender. But for the men who fought its battles, the war still goes on, played out in memories of those brief, harsh days.

At the age of fifteen, Denzil Connick became a Junior Soldier. Two years later, in 1974, he joined the Parachute Regiment. He lost his right leg to an Argentinian shell two days after the battle for Mount Longdon. Today, he is Secretary of the South Atlantic Medal Association, an organization that seeks to maintain links between veterans of the South Atlantic campaign and to strengthen friendship with the people of the Falkland Islands.

In 1982, at the time the Falklands War began, I was a lance-corporal with Support Company, 3rd Battalion, Parachute Regiment (3 Para). We were stationed at Tidworth, on the Salisbury Plains, in Wiltshire. And then the war came along, just a shot right out of the blue. Nobody warned us that this sort of thing was going to brew up, and it was a total shock for most people.

At the time, 3 Para was on spearhead, which is 24-hour standby alert. We were prepared to be placed anywhere in the world at a moment's notice. All our intelligence briefings up to that point focused on the possibility of things going wrong in Northern Ireland, the Middle East – the typical hotspots. So when the signal came through that we were to mobilize to go to the Falklands, half of us didn't even know where the Falklands were. We thought, what are the Argentinians doing invading some remote islands off the north of Scotland? It was sort of an insignificant place to most of us at the time.

And then, we were there…

June 11 was the day I first met the war, in its rawest sense. That's when we went in to full-blooded action. When you're actually going into battle, the fear is so intense it's almost unbearable. You can barely breathe. Your heart's pumping, your adrenaline's flowing. Your head feels like it's going to burst. And there's a natural instinct to do one of three things: either curl up in a ball in the fetal position, or run – straight ahead, or back the way you've come. Flight or fight. And because of our training, we all fought.

As we advanced for the assault, it was almost as if I turned into a robot. Your training overtakes you, and you become almost automated, but your feelings are still there. Your body is doing things, but your brain is somewhere else. My legs felt as if they had just been filled up with lead. It was almost like a physical and mental effort to actually keep advancing toward this hell that I was walking into.

We planned to carry out a silent night assault on Mount Longdon – there had been no artillery bombardment to soften up the enemy position before we went in, which is the normal procedure. A silent assault gives you only one advantage: surprise. We aimed to be literally in and amongst them before they knew what was happening. And we hoped to be able to bayonet them in their sleeping bags before they even had a chance to grab a weapon.

What actually transpired was that one of our guys stepped on an anti-personnel mine, vital minutes before we got onto the first of the Argentinian trenches. The explosion and screams alerted them, obviously. The battle started in earnest just moments later. So it was a hand-to-hand bayonet assault with a prepared and alert enemy, which put us almost straight away in a disadvantage. They were in a defensive position, in trenches and bunkers. We were fighting up a mountain towards all these well-defended positions. We had to fight for twelve hours, hand to hand, trench to trench – the old fashioned way.

And that's what we did: twelve hours of hell. Never mind all this "God, Queen and Country" stuff. You were fighting for your mates. That's what it boiled down to. We had the sea behind us, we had the sea in front of us. There was nowhere for us to go except forward and win the battle. And we did it. Against the odds we actually did it. We overran their position and killed or captured dozens of them in the process. We'd lost something like twenty-two guys up to that point, dead, but three times more than that were wounded. We had killed something like fifty-odd of them and three times that – maybe 150 to 200 – were wounded, and a hell of a lot more became POWs.

I never actually thought in terms of *people.* I kept that out. You can't really think or look into a person's eyes as you are killing them, because you're just torturing yourself. What you're doing is taking something out of the way that's in your way – and so it's got to be a thing. Some*thing* is stopping you from getting from point A to point B. So you get rid of it. In your mind, you can't think too much about the fact that it's a human being that you are facing. You almost desensitize yourself to get on with the job.

I remember clearly something that happened right after that battle for Mount Longdon. There I was, surrounded by corpses, and I actually made a cup of tea using a body to shield my little cooker from the wind. I was doing it without the slightest bit of morbidness, and there was no disrespect to the Argentinian corpse lying next to me; it was still doing a job – it was actually stopping the wind blowing my little cooker out while I was making a cup of tea. I just turned him up a little on his side and made a bit of wind break out of the corpse. I did it without even thinking. I mean, it sounds horrible, and it probably is, but at the time it wasn't. I just did it out of necessity. I even had a quick look around to see if the guy had some water I could use. He didn't need it anymore, you know.

And then we had to set about collecting the bodies of our own friends. That is one of the most unpleasant things I've ever done – carrying the stiffened corpses of friends down to a place of burial. It starts to get home to you then, the beginnings of the consequences of the previous two days of battle. But just the beginnings of it. You can't allow it to overtake you, because you've still got a job to do. You still have battles ahead. You can't allow these things to creep in, because they could spoil your effectiveness for future battles. So you put it out of harm's way.

On June 13, two days after 3 Para and the Royal Marines 42 Commando and 45 Commando took Mount Longdon, Denzil's

war ended. Heading off the captured mountain, now subject to Argentinian defensive fire, he stopped to chat with two fellow soldiers. None of them heard the shell coming until it was too late. It exploded right where the three were standing. Denzil lost his right leg. The other two men were killed.

BIG BOYS DON'T CRY. That's what Vince Bramley was taught as a child. War taught him otherwise:

I remember one time when I was about six years old, some other kid stood outside our house and kept calling me names. My mum kicked me out of the house to go and face him. I had to fight this kid, and my mum stood there and watched! I won, but it was a hard lesson to learn. That's the way we were as kids, and that's the way I was during my years in the Army. I'm not now. Some lessons I learned the hard way. Going to war played a big part in that.

At the time the Argentinians invaded the Falkland Islands, I was a machine gunner with B Company, 3 Para. We had just returned to England after training and a hearts-and-minds operation (going around handing out sweets and toothbrushes, for the most part) in Oman. I was still suntanned from the desert, and my head was still full of desert warfare training. Of course, once it became quite clear we would be going to the Falklands, we were in a completely different frame of mind, and our training shifted focus rapidly. However, due to the urgency of getting the Task Force moving, we were literally bundled onto a ship and set sail very quickly. We didn't really have time to think about where we were going, what we were doing. It wasn't until we were onboard ship that we realized we were heading into near-arctic conditions. We were going to an area largely unexplored by the British; the Falklands' coastal lines had only been mapped the year before by a Marine officer who was keen on sailing.

The rest is history, getting down there. I mean, there were different ways they could have stopped the war, had they wanted to. But I really believe, as I stand here sixteen years later, that Margaret Thatcher wanted this war, and I believe the military wanted it as well. From their point of view, we needed a campaign to test our training program. Sure, we had Northern Ireland, but it couldn't compare to conventional warfare, clearing bunkers and trenches and so on. Added to that, our kit and equipment were becoming outdated, and the military wanted to see how it would hold up.

At the time, none of us thought too much about the whys and wherefores. It's only as you become more educated afterwards that you realize the war wasn't just a bit of prestige and political argument. There's more to it than meets the eye, which is only just coming out now: the potential bonanza of oil down there, for example. But we weren't thinking down those lines at all.

Up until the Falklands, I saw myself as the rough-tough soldier I had joined the Army to be. But the war taught me otherwise. I saw some of my best friends completely blown to pieces, and I had to put them into body bags and carry them down the hill. I saw a seventeen-year-old kid get killed right in front of me. He wasn't old enough to vote or drink in a pub or go to see an X-rated film in a cinema, but he was good enough to die at the age of seventeen. Another kid died during our attack on Mount Longdon. He turned eighteen the night he was killed.

I remember that battle well. Mount Longdon was about ten miles outside of Stanley, and from the summit you could actually see the town and the cars going up and down. Longdon is classed as a mountain, but it's only 700 meters high; really, it's just a very big hill. But it has a very steep western slope. Once you get to the summit, however, it peters out into a ridge about a kilometer long. That ridge runs into what we called Wireless Ridge, which goes round to the bay of Stanley.

South and a little to the west of Longdon are Two Sisters and Mount Harriet. The plan was to take these three mountains in one night, the night of June 11. My battalion, 3 Para, was to take Mount Longdon, the Royal Marines 45 Commando would take Two Sisters, and 42 Commando would take Mount Harriet.

At the "start line" at the base of the mountain, our commanding officer gave the order to fix bayonets, and then the boys of B Company headed up. Once they were fired at, it was every man for himself. Although each corporal knew exactly what to do to take the objective, once the battle started communications between the corporal and his privates went out the window. But the soldiers quickly formed two- or three-man teams, and without any orders from their corporals or their superiors, they started to take out the bunkers by grenade, bayonet, or rifle fire.

While the first wave of soldiers climbed into battle, I was giving traversing support fire on my machine gun. But because our men got to the summit within a couple of hours, the machine guns couldn't really be used any further. I would have been shooting down my own men. So our orders were quickly changed. We picked up our kit, walked through a mine field, went straight up the slope that B Company had already climbed, and caught up with the tail of B Company, who were still fighting through the bunker defenses.

Once we'd taken the summit – there was a sort of natural bowl there, and in it the Argentinians had established their main positions – we set up the machine guns and fired across the summit, taking out the enemy positions on the slopes opposite. Now, this was all at night. The battle began at midnight, and there's about sixteen hours of darkness in the Falklands during the winter. By the time we knew we'd achieved our objective, we were getting artillery support and naval gun support fire, which was coming from the ships fifteen or twenty miles out. Their rockets were flying over our heads and going in.

But for the most part, this was combat at very close quarters, hand to hand, eye to eye, very bloody stuff. On the rifle range, we'd always been taught, "Target will fall when hit." And that's the way we had to think. You couldn't think of the target as another human being. We kept it simple: target will fall, achieve the objective, kill and move on, get this whole thing out of the way.

It wasn't until daylight, when I ran into the bowl on the summit and saw the number of dead people there, including my own friends and colleagues, that the shock hit me. Nobody touched me, but it was as if somebody had punched me in the stomach. And I just went into a state of shock. It's hard to describe the feeling, but during battle your adrenaline is running so fast and so high you think everything's happening in slow motion. Your brain is going twenty times faster than it normally would, because you're so alert. Only later does reality hit you.

I remember looking around at some of my friends who had survived as well and were in this bowl, and I hadn't realized until then that I wasn't the only one crying. And there were Argentines who'd been taken prisoner, and they were crying as well. I think all of us were shocked at the extent of what we'd done to each other. And then you begin to realize you're not the rough, tough British paratrooper that the program of training had made you out to be. You realize you're human, and you have human feelings, and that the men beside you are no different. The one thing that united all of us bundled together up on that mountain – both Brits and Argentines – was that we were all very upset about the whole fucking mess we were in.

KEN LUKOWIAK SERVED in the Falklands with the Parachute Regiment's 2nd Battalion and took part in the battle for Goose Green. He remembers:

When the colonel came into the gym at the barracks in Aldershot and told us we were slated to go, my first reaction was a happy, joyous one. Everyone was all gung-ho, rah, rah, rah. If we'd been left behind it would have been unbearable; we were really happy to go. It certainly had the feel of a great adventure. The prospect of having to kill people didn't bother me at all – it didn't come into the equation. This was the Parachute Regiment. We were professional soldiers, and that's what professional soldiers do – kill people. You don't need to sit down and have a think about that too much. It stands to reason.

But it was something of a journey from there to the point when the first shell came in at Goose Green. From that moment on, you're zapped into another place: it's like tripping on hallucinogens – another state of consciousness. For me, it was as if the Falklands War took place in another dimension. Everything was turned upside down. The things you thought were important and that you worried about – your mortgage, or whatever – didn't matter two hoots. Suddenly you've stepped into an environment where there are only two types of people: the ones that want to kill you and the ones that want to keep you alive. And it's a very pure living. And afterwards, you remember the details, what it looked like. But you can't really remember the feeling.

The battle for Goose Green Settlement was fought in the dark, across rough terrain. The settlement at Darwin, at the mouth of an inlet off Choiseul Sound, was our first objective. I was assigned to putting down covering fire from across the inlet, which left me a bit out of the thick of things. So at that point the war was just bangs and booms and things exploding in the dark across the water. But as the sky grew lighter we pressed on towards Goose Green. We were walking across fields, and I could make out the shapes of bodies lying about. My first thought was that they were department-store dummies. I really thought, fuck, isn't that cool? A truck must have come along and dropped them off. Then we got up to them, and I could see that they weren't dummies, they were dead people. Dead.

Later, I saw more dead people, people that I knew. I watched some of them die. At the time, it didn't seem significant, really; I was just in a different state of mind. And that's the thing I'll never be able to explain…

I know a lot of people, myself included, who had trouble dealing with the war once it was all over. But if I try to look at it honestly, I have to ask myself, "Am I using the war as an excuse for mistakes I've made in life?" I don't want excuses. At the end of the day, we men of the British Task Force were all professional soldiers. No one forced me or any of my mates to join the Army. It wasn't like Vietnam, where you were minding your own business at home and a letter came through the door, and next thing you knew you were in Saigon or wherever. We have to take our share of the responsibility, because we were the ones who walked into the careers office and signed the dotted line.

Now the bit I can't get a grip on is why we cheered when we first heard we were going to war. I cheered. I thought it was fucking great! And had they said, "You've got to stay behind," I would have been gutted. I can't work that out. I can't rationalize that. I wasn't sitting there thinking, I'm gonna be able to kill some people! There was none of that. I don't think any of us thought like that. There were those who said, "Good, I've waited fifteen years to do my job." That wasn't my thought, but still I was happy when the chance to go came along. I cheered. But guess what? I wouldn't cheer now.

Ten years after the war ended, Ken began in earnest to sort through his experiences, putting his thoughts on paper as a way to come to terms with his tangled emotions. The result was *A Soldier's Song,* a book that transpired more by accident than design, and for which the *Mail on Sunday* dubbed him "a back-street Wilfred Owen." The closing chapter reads:

Since the Falklands War I have often gone into British Legion Clubs. On the wall in most of them there is a plaque. On the plaque these words are inscribed:

> "At the going down of the sun, and in the morning
> We shall remember them."

For me these words are a lie. When I awake each day my mind does not recall the dead that I once knew. I am usually occupied with thoughts of how I am going to pay the outstanding bills I have pinned to my kitchen wall or what the day before me will hold. At night as I lay my head on my pillow I seldom recall the dead that I once knew. I am normally trying to understand the confusion I have created around my life over the past nine years, as I watch vivid mental replays of my past insanities.

At the end of the evening, in the rooms with plaques on the walls, all the old soldiers stand to attention and, in the same way an actor recites an author's lines, they say aloud the words that are inscribed on the plaques. I look around the rooms at the faces of the old soldiers and wonder if they might also be lying? My own hypocrisy disturbs me, for I know that as long as the majority of us continue to act out the plays that have been written for us by the politicians, their priests and the men of this world who control the money, then we shall never be able to put an end to the horrors of war.

I do remember the dead that I once knew, the dead on both sides of the war that I fought in. But I remember them not with the part of me that sits drunk amongst old soldiers and brags of the glories of war, but with the part of me that has seen war, knows of its true horror, the stupidity of it, and still feels an inner unrest for having witnessed it.

I do remember the dead that I once knew, but not with the part of me that loudly shouts, "What about the filthy spick fucker that the sarge blew away?" but with the part of me that when in solitude, whispers a prayer for the one-eyed, dying boy from the Argentine.

I do remember the dead that I once knew, but not with the part of me that speaks of brave young men who charged to a heroic death, a death for the Queen, a death for a just cause, a death for their country. But with the part of me that tells of the young men who had their flesh ripped and punctured by flying metal, young men who screamed and died in agony, young men who prayed for mercy, prayed for mercy to a God they had never accepted before.

You see, I do remember the dead that I once knew – sometimes.

I would like to change the inscriptions on the plaques. I would like to be able to give myself a chance of being truthful. I would write these words instead:

"Sometimes – when we are able to find our true selves – we shall remember them."[17]

The Gulf War

ONE NIGHT IN mid-January 1991, I saw on television what at first appeared to be a highly realistic video game, with jets screaming in and out of view, bombs exploding in tidy, white puffballs, and an excited commentator providing a blow-by-blow account of the action.

It took me several seconds to realize that I was watching the launching of Desert Storm, the U.S. air war against Iraq. What struck me more than anything else was the commentator's tone, which was one of excitement and pride in the display of U.S. technical prowess. The names of the aircraft and bombs used in the attack were described in glittering detail, like a roll call of American military accomplishments. But there were no explanations of the bombs' effects, no images of what was happening down there beneath the tidy,

white puffballs. And there was no discussion whatsoever on the cause of the war itself.

On the surface, it seemed to be about oil and the urgent need to stop a brutal dictator. But there was more: a kind of panic had set in at the collapse of the Soviet Union and the abrupt end of the Cold War. No one denied that the last serious threat to American national security had been removed, but the military budget was still enormous, and the defense industry seriously bloated. To begin to dismantle it all would mean the loss of hundreds of thousands of jobs.

There were, of course, political ends as well. In October 1990, the *Washington Post* had reported: "Some observers in [Bush's] own party worry that the President will be forced to initiate combat to prevent further erosion of his support at home." Elizabeth Drew, writing for the *New Yorker*, quoted Bush's aide John Sununu, who said that a "short successful war would be pure political gold for the President and would guarantee his reelection."

Finally, there was what Bush himself had dubbed the "Vietnam syndrome," an American boil which had to be lanced. The invasion of Panama had been too small and too inconclusive to weaken the public's abhorrence, since Vietnam, of foreign military interventions. And yet a war against Iraq could not drag on indefinitely; it had to be brief enough to preclude the development of a significant antiwar movement.

Bush's war worked. The punishing air strikes against Iraq brought about an immediate surge in Bush's popularity, and the defense industry was satisfied. A cheering crowd of executives at the opening of the Fifth Annual Defense Contracting

Workshop chorused, "Thank you, Saddam Hussein!" And as the bombing drew to a close, a beaming Bush reflected, "The specter of Vietnam has been buried forever."

The Pentagon, it seems, had learned the lessons of Vietnam well. If Studs Terkel dubbed his oral history of World War II the "Good War," then the attack on Iraq could aptly be called the "Clean War." Restricted press coverage of the war on the ground helped create the perception of a quick, decisive victory at low cost to human life. Air strikes hit their targets with "surgical" precision, and sophisticated, laser-guided "smart bombs" added a mystique of high-tech thrill and cool vanity. U.S. casualties were minimal.

Though the war failed to win reelection for President Bush, he retired gracefully. In a 1996 television interview with the BBC's David Frost, he reflected, "The mission was to end the aggression...I don't think war is immoral...I think history will say we did the right thing." But what really happened in Iraq is a story without an ending. And it is one that only the Iraqi people and veterans of the Gulf War can truly tell.

Ask Paul Sullivan, director of the National Gulf War Resource Center and combat veteran of the ground war in Iraq, and he'll describe a war far removed from CNN's depiction of glistening fighter jets and little white puffs of smoke:

When you see the battlefield littered with dead bodies as far as you can see and there's smoke swirling around, and the smell from the dead bodies, the ammunition, the fuel, the explosions, it's very overpowering. It's very sickening. You're offered very, very little opportunity during wartime for any type of introspection. You are so geared toward killing and surviving that any other type of emotion, as well as any type of reaction to what you're seeing, is deeply suppressed. I think some of that comes about as a result of the training. In the military you are focused on "the mission," as they say…

You see it with the soldiers. They'll set their lips, tighten their eyes, grimace, and just keep looking forward. They can drive through miles and miles and miles of charred trucks, tanks, blown-up buildings, pieces of arms, pieces of legs every which way. Folks may remember what they called the "highway of death" that led from Kuwait up to Basra. That had to be one of the most hideous, grotesque, disgusting abominations that I've ever witnessed in my entire life. And it's a result of the lies. It starts when people say, "That's mine. You can't have it," or, "I'm better than you." That's where it starts…

In the Gulf War, we were the world's largest, fastest, most well-trained and well-equipped firing squad that ever existed. We systematically cut off Iraq's water, electricity, telephones, you name it, and completely immobilized the whole country. It was like a boxing match in which you blindfold your opponent first and then tie his hands behind his back and then turn out the lights. They couldn't see us and we just sat there and pummeled them for thirty days. And then we came in with the thousands upon thousands of tanks, all lined up side by side, across hundreds of miles of desert, every tank within eyesight of the next tank, and swept across the desert like fire ants, killing, blowing up, destroying, crushing, demolishing, and blasting every single thing in front of us. It was a "scorched earth" campaign, and when we were done there was nothing left except the charred remains: tanks, cars, bodies and body parts strewn all over the desert. That was the Gulf War. It's not like some smart bomb just went down Highway 1 and hung left at the stoplight and went down Highway 5 and then went smoothly down a smokestack to destroy some evil Arabs who were plotting to destroy American women and children. That's just not true…

ANNE SELBY WASN'T always a soldier, and she certainly isn't anymore. Before her Territorial Army days, which merged into an Army career and consequent Gulf War service, she worked as a professional singer, trained to sing opera. Today, her voice is a dry croak. In conversation, she speaks in thirty-second lurches, then stops to rasp and wheeze, coughing phlegm into a wadded-up bath towel.

She's seen the specialists, and left them baffled. Her GP says her lungs look like they've been through a house fire. But Anne isn't particularly surprised: she knows she inhaled plenty of toxins from blazing oil wells during the Gulf War, and guesses she was subject to chemical contamination as well – to say nothing of the medicinal cocktails injected into her bloodstream by British Army nurses. Her symptoms match those of countless other veterans suffering from what has come to be known as "Gulf War Syndrome." To Anne, one thing seems certain: war changed the pattern of her life forever. Though she spent her days in the Gulf in relative safety, as a Staff Sergeant attached to a hospital base's laundry platoon in Saudi Arabia, she by no means escaped the effects of the war.

The Gulf War is the only war I've ever been in, and it left me feeling dirty. Just plain dirty. I still don't feel entirely clean. It's taken me years to come to terms with everything in my head. It sounds crazy, but when I returned from the war, I had a lot of anger, which then turned to guilt. I was made to feel very guilty because I returned; we weren't expected to come back. You know, they made elaborate preparations for massive casualties that never came, so it was almost as if we let them down, somehow, by not dying. Before we left for the Gulf, an officer actually told me to my face that we were not expected back alive; they were shipping hundreds of coffins to accommodate our return.

There's only one word to describe what I saw of the war: unreality. The place where we lived felt like a Butlin's holiday camp: you know, those horrendous cheapo camping grounds for families – that's what it reminded me of.

There would be these little girls in bikinis, with their tans and their little flat bellies, stretched out on sun lounges. The only way you knew they were soldiers was because they had their NBC (Nuclear, Biological, and Chemical) kit piled beside them. Others strutted around in little halter tops and shorts, fiddling with their hairdos. Even during SCUD alerts, invariably there were those who wouldn't bother to suit up in their protective kit. The war seemed so remote, so far removed from us.

I'd had three weeks of Gulf War training, three weeks of having the importance of "kitting up" drummed into my head – in the event of a missile attack, you get your kit on as quickly as possible; otherwise you're dead. I'd been led to believe I was going to war, so I hadn't taken any civilian kit with me. And then I get to Saudi Arabia, and the whole thing is just totally bizarre.

There was a swimming pool at one of the camps occupied by U.S. troops. It was theoretically off limits, but that sure didn't stop the Americans. They more or less annexed the swimming pool, and from then on there seemed to be a running pool party going all the time. I'd be working a twelve-hour shift, and I'd hear stories about barbecues around the swimming pool! And I couldn't hardly help but think, Hang on, what kind of a war is this? We're in two different wars here! It was unreal.

Nearly everyone, it seemed, had brought along a camera, and they wandered around taking photos, which only added to the impression of being at a holiday camp. Bikinis and cameras and people lying around in the goddamn sun...

Of course, where I was I certainly didn't see much of the slaughter. But I know people who did. I spoke to a young British soldier not long ago who had joined other Allied troops in burying Iraqi soldiers alive in their trenches. He was very traumatized by the ordeal. They just buried them alive! And those weren't by any stretch of the imagination Saddam's crack troops. By the time our lot got that far he'd withdrawn his best soldiers. Those that were left were conscripts, badly equipped, half starved – they were nobodies; they weren't a threat to anybody!

The POWs that surrendered to our troops were a lice-ridden bunch. They were malnourished and had scabies. In many cases, their officers had deserted them. We saw them when they brought them back to the hospitals. Those men couldn't have blown their way out of a paper bag, never mind killed anybody. They were kids – and I mean young! You know, Saddam had just conscripted everybody. And it was people like these that our soldiers were ordered to hound across the desert.

Perhaps Wendell Berry, a man who has written passionately about the demise of American culture for nearly four decades, says it best:

[The Gulf War] was said to be "about peace." So have they all been said to be...But peace is not the result of war, any more than love is the result of hate or generosity the result of greed...

This was a war to bring about a "new world order."...In fact, this war produced not order but disorder probably greater than the disorder with which it began. We have by no means shown that disorder can be put in order by means of suffering, death, and destruction.[18]

The boys fight
the wars of men.

Herman Melville

The art of military recruiting is an ancient one. Militaries throughout history have used the darker side of human nature to their advantage and have long understood that "the passions that are to be kindled in war must already be inherent in the people."[1]

In time of war, those passions are more base, more immediate. Young men are driven by hate and fear, not only of the "enemy," but of the reactions of friends and family and society, should they fail to conform. Many would rather die than risk disapproval and loss of respect; would rather have their blood ebb away on distant sands than turn against the tide of common emotion and expectation. As Vietnam veteran and author Tim O'Brien observed, "The soldier's greatest fear [is the] fear of blushing. Men killed, and died, because they were embarrassed not to... They were too frightened to be cowards."[2]

In times of so-called peace, these "inherent passions" are less immediate but can be exploited nonetheless. Ambition, pride, elitism, and wanderlust; the quest for physical prowess, power, and control – all these are engendered by familiar slogans like the U.S. Army's "Be all you can be" or Britain's "Army soldier: be the best."

Anna Simons, an anthropologist at UCLA, observes that young recruits shed much more than pounds during boot camp. She notes that drill instructors place paramount emphasis on stripping away critical and independent thought, and end up recalibrating graduates to a dangerous level of group-think. She recounts statements by drill instructors such as: "Before we leave my island, we will be thinking and breathing exactly alike," and "Nobody's an individual, understand?" This drive for conformity and homogeneity goes to such excessive lengths that sentences beginning with the first person "I" are strictly forbidden. Surprisingly, she concludes that such indoctrination might be beneficial for dissolute youth, and endorses a program of compulsory national service which includes the discipline of boot camp.[3]

But while the military makes much of team cohesion and plays heavily on the human desire to belong, its ultimate purpose is the calculated killing of other human beings. Buried under the lofty goals of self-sacrifice and self-denial for the good of the whole, the noble end of training is nothing other than death.

We make much of cults whose leaders deceive naïve followers and lead them down paths of death and destruction; the ire of the American public is kindled at the very mention of Jonestown, Waco, or Heaven's Gate. But what would people say if they were to learn of an organization that entices hundreds of thousands of teenagers into its ranks each year, offering money and adventure, a new family, a new identity, and a promise for membership in "one of the most elite organizations ever created"? And what if they heard that this organization makes it illegal to leave, and imprisons or shoots members who try? What if they were told that the leaders of this organization systematically corrupt the morals of its new converts, teach them the dehumanization of whole nations and classes of people, take them to foreign countries, put weapons in their hands, and force them to kill? Wouldn't such an organization be perceived as one of the biggest cults of all time?...

THE MILITARY ENTICES YOUTH with all sorts of promises: higher education, travel, excitement, camaraderie, purpose, and honor. But the truth is that even when these promises do materialize, they can come with a hefty price tag. Young people – especially those looking for a ticket out of poverty, family strife, inferior education, and disenfranchisement from middle-class opportunities – are very often disappointed. Jim Murphy, a Vietnam veteran and dean at West Side High School in Manhattan, explains:

It's an American tradition that the military is viewed as a career option for our kids. My kids are especially susceptible to this. The high school where I teach is about 50 percent Latino and 50 percent African-American. Most of my kids come from Washington Heights and Kings Bridge, or Harlem. Ninety-six percent of the kids in my high school are eligible for the free-lunch program. That's a very clear economic indicator; it says that they are living in borderline poverty. Last year I lost three students, killed during the summer. The year before that I lost six kids; the year before that there was one. Every year I walk back into school in September, and someone's been shot. That's the environment in which a lot of my kids live, day in and day out.

There was a very interesting study done on the demographics of my kids' neighborhoods as compared to Westchester County. They asked kids from Harlem if they'd borrow $10,000 to go to school and about 80 percent said no, they wouldn't borrow that much money. Then they asked the same number of kids in Westchester if they'd borrow $100,000 to go to school and 75 percent said yes, they would. So my kids don't have a sense of success in their future. And when a recruiter comes to school and says, "We can work out a deal where you're going to earn $40,000 for college," that's really a dream.

He's offering a career, he's offering money for college, yet when my kids go into the military they have a failure rate that's pretty high. A lot of them just aren't going to accept a racist comment from a drill instructor. They are going to react to it. And the military itself, the basic training, is based on breaking people down. I would imagine the average drill instructor probably insults everyone about equally, but my kids aren't going to take that. So they leave, and they don't get the $40,000, and because they've "dropped out," now they are going to have a difficult time getting even a simple civil-service job.

Not surprisingly, the whole subject of war is neatly excised from military advertising. Instead, a U.S. Marine Corps homepage advertisement focuses on flattering images and the mysterious wonders of boot camp:

From the very first haircut to the last drill evaluation, we take each step of boot camp with one thing in mind: pride. We want you to walk away proud to wear the title United States Marine.

Sure, we demand a lot of you, from tough physical training to long hours in the classroom studying Marine Corps history and traditions. You'll learn the applications of strategy, tactics, and how to handle the challenges only The Few can.

With every exercise you'll become stronger. Because we want you to be the best there is. We want you to develop that spirit of "nothing's impossible." The unwavering self-confidence that comes from being one of the best.

We're inspiring you, motivating you, while you show us you've got the heart and smarts it takes to become a Marine.

There are times you'll wonder why you're doing it, but you won't wonder for long. Because twelve short weeks later comes Graduation Day. The day you realize very few people can accomplish what you have.

After you receive your basic uniform issue and finish your physical examination and administrative processing, you'll meet the individual who's going to make these next twelve weeks worth every ounce of sweat and determination you put into it. He's your drill instructor, better known as your D.I.

Even though there are times you'll wonder whose side he's on, you'll learn he's the one man you'll want by your side. He'll teach you everything you need to know. And he'll never demand anything of you he wouldn't demand of himself.

He'll show you how to fire a rifle. How to conquer the obstacle course. How to get yourself in shape and stay that way. How to build mental and physical confidence. How to respond to orders.

Throughout boot camp he'll be there. Pushing you and giving you the discipline you need to be your best.

But more importantly, he'll show you what it means to be a team player. To make a group of men function as one. In short, he'll teach you what it means to be part of one of the most elite military organizations ever created, the United States Marine Corps.

On the highway not far from my home in upstate New York is a large billboard with a picture of a U.S. Marine in full uniform, standing ramrod straight and at attention, with a firmly grasped sword lightly touching the brim of his gold-braided hat. The caption reads simply: "The Change Is Forever." Having gone through boot camp in the Marines myself, I can attest to the truth of that statement, though not in the sense intended by the advertiser. Rather, I must side with Vietnam veteran Gerald McCarthy, now a professor at California's Thomas Aquinas College, who says:

Going to boot camp *will* change you forever. I wouldn't let my son go to boot camp. Forget going to war – I wouldn't even let him go to boot camp. I wouldn't go back to boot camp for all the money you could give me, because I think it can change you psychologically forever. Why would you want to lose your youth? There are so many ways to spend it.

Though they number far fewer than their male counterparts, women in the military face the same kind of deception as do the men. Carol Picou, who served in the Middle East during the Gulf War, now suffers from what is probably the combined effects of toxic medications and radiation from depleted uranium – so called Gulf War Syndrome: risks about which she was never warned, and which the Pentagon still denies. She wishes she could be allowed to place her own advertisement alongside the recruitment billboards.

I would love to make a commercial now. When I joined the Army it was, "Be all you can be." So take us and show us in basic training, firing our weapons, climbing mountains, rappelling, doing all these wonderful things the Army teaches you to do, and then show us now, with our crippled bones, our incontinence. Take all of us in our wheelchairs, missing our arms and legs, and dying of cancer and brain tumors. Take our graves, and put *that* on a commercial.

"Be all you can be!" Now, what *can* we be? Would they let that be aired? Of course not. And who would make my commercial? Nobody.

Military advertising, such as a U.S. Marine Corps letter to high school seniors, is less than sincere when it encourages young people to "get all the information you can, from as many sources as possible." They would never, for example, voluntarily tell

anyone the story of Vietnam Veterans of America founder Bobby Muller, who said:

I went into the Student Union Building, and there was a Marine officer... He looked very sharp: he had his dress blues on, and he had the old crimson stripe down the side of his trousers. I said, "That looks good! I'm going to be a Marine."

Right there, in that sentence, is really the tragedy of my life, as I view it. The tragedy of my life was not being shot in Vietnam; the tragedy in my life is one that has been shared by all too many Americans, and is still being shared today. For me, knowledge of the fact that my government had seen fit to involve us militarily in Vietnam was sufficient for me. *I never asked the reason why.* I just took it on blind faith that my government knew a hell of a lot more than I ever could, and that they must be right. My opinion has changed since then.[4]

Perhaps one of the most tragic aspects of military advertising is the way in which it appeals to those young men and women whose lives have been shattered by dysfunctional families and communities. Such advertisements go to great lengths to replace the natural yearning for family with hopes for camaraderie and acceptance in the military. For example, the letter to high school seniors asks, "Who will give me a sense of belonging?" It then promises to make them part of a team, a "group of men who function as one."

David Harvey, a British Special-Ops veteran, served in Italy, Egypt, Germany, China, Kenya, and Hong Kong. He was an orphan when he enlisted at age sixteen, and he joined the military in search of a better life.

The reason I joined was that I didn't have a family. I'd left school at thirteen, and I had a number of mundane jobs that were going nowhere. The country was at war, and there were a lot of military personnel about. I guess I had the idea that maybe I'd be okay in the services. I didn't look at it as a career. I looked at it as a way out. There was nothing for me in civilian life, not at this stage anyway.

In the service, we lived quite a good life. We had plenty of spare time, plenty of sports, and I couldn't see anything wrong with it at all. I had a wage, I had food, I had clothing, and I had comradeship, which was very much encouraged in the services. I thought it was a good life.

Now, at age seventy, he reflects on the years that led to his retirement from the British military at the age of thirty:

I looked back on my life in the service and realized it had done me no favors whatsoever. I had no guidance. I didn't have parents who were able to warn me of pitfalls. And I guess I was something of a mess when I first came out…

It was all a grave mistake. I don't have anything to look back on, apart from the comradeship and the fellowship of similar servicemen. The rest, the things that I was asked to do and did willingly – had I known what I know today, had I been a wiser person at the time, I would never, ever have done them.

I knew that what I was doing wasn't right already before I left. I didn't actually have to wait until my time in the service was over to come to that realization. But of course, as you get older, you reflect more deeply on your life, and your regrets become stronger. It isn't something that fades away. You just keep saying to yourself, "I wish I hadn't." But I have. I was young and naïve, and now I'm older and wiser.

Retired Navy Rear Admiral Eugene Carroll serves as deputy director of the Center for Defense Information in Washington, DC. A man of impeccable credentials, he is well-qualified to pass judgment on the institution he has served for over forty years (he served on destroyers in the Pacific and in Vietnam, and in Cold War Europe with General Haig). When asked what he would say to an idealistic young person contemplating a career in the military to make the world a better place, he responded:

First, I'd disabuse him of the notion that the military makes anything better. The military exists to kill and destroy. Now you can rationalize it, and you can justify it on the grounds that, well, there's great evil out there that we must redress, but that involves killing and destroying. And everybody is the loser, humanity is the loser. So I would certainly disabuse him of any idea that we're going to be a force for uplifting people and the quality of life in this world. There are all sorts of practical and very needful ways to help improve life, and the military isn't one of them.

Retired Vice-Admiral Jack Shanahan is director of the Center for Defense Information. He enlisted in the Navy prior to World War II, served in the Pacific off the coast of Korea, and commanded a number of tours in the Tonkin Gulf during Vietnam. He also spent a year in-country as Commander of the Coastal Surveillance and Interdiction Force head-quartered in Cam Ranh Bay. Shanahan once commanded the U.S. Second Fleet and NATO Strike Fleet Atlantic, the guided missile destroyer *Cochrane* and the destroyer escort *Evans,* and six other destroyers and missile ships. Though he upholds

his belief in a strong national defense, he is committed to exposing gross waste, senselessness, and mismanagement of military matériel and personnel. He, too, has some remarkable insights to offer into the long-term consequences of military training:

Now that we've created a professional military force, we expect people to be signing up for a career. In other words, they should know when they're going in that this is what they're going to do. Instead of being a lawyer or doctor or writer, they're going to be a soldier or sailor or a Marine. And they're probably going to do that for twenty-five years. And then they're going to leave the service. And, quite frankly, there isn't a lot of gainful employment out there after a guy has spent twenty-five or thirty years in the military. What is he qualified to do on the outside? Not very much, and that is a significant problem.

To young people contemplating a military career, I'd say, "Dig. Make sure." Because if you're going to go down that path, you ought to know exactly where you're going, and not just accept the word of the guy who's trying to get your name on the bottom line. You want to make sure that you look under all the rocks.

People need to know that the needs of the service will come first. So if the needs of the service change, even after you've signed, all bets are off. You want to make sure that you look very carefully at what you're talking about, because nothing is certain. Go to other people, to other sources. And I don't mean other military sources: talk to your peers, to your teachers, to your high school guidance counselors.

DAVE HIBBS JOINED the British Army at the age of sixteen, in 1966, lured by the prospect of learning skills that would later put him ahead in civilian life. He remembers how quickly illusions

of career advancement were displaced by the harsh reality of basic training:

I grew up in a small mining village in North Nottinghamshire. My father drove an ambulance for the local coal mine, and after seeing the horrific casualties he sometimes had to transport, he said that I should not go "down the pit." In fact, he threatened to throw me out of the house if I went to work for the Coal Board. As mining was just about the only job available in the village, I had to really think about a career.

I was not a brilliant enough scholar to make education my "escape route," and I left school with one "O" level and a handful of low-grade CSEs.

I drifted into the summer holidays just after my sixteenth birthday with not a clue what to do. My father and mother both worried over my future, and my father made enquiries locally. One day he announced that he'd spoken to a lady who lived in the village who recommended a career as a physiotherapist! I hadn't a clue what PTs did, so I was sent off to this lady to talk to her. She told me that you had to have at least two "O" levels to be accepted for training – bad luck, as I only had one. Then she told me that the British Army would accept men for this training with only one. Wow! I was there.

At this point I still didn't have a clue what a PT did for a living, but I was obsessed with wearing a white coat all day and being something vaguely medical. I told my parents and they thought it was a fine idea, so I went off to our local market town one day to "find out."

The recruiting officer was a well-educated sergeant who immediately put me at ease. "What would you like to do in the Army, son?" he asked with a smile, and I told him of my dream of becoming a PT. He said that he could certainly offer me a place in the Medical Corps, but once in I would have to take further tests and be interviewed by the Corps Training Officer. He assured me that this would not be difficult for a boy who was obviously so intelligent. Little did I know that all recruiters had a quota to fill each week for different regiments or corps. Mostly they were looking for infantry or cooks, and the recruiter asked me if I fancied the Grenadier Guards! Luckily my poor eyesight saved me from that. I was asked to come back in a week to take a medical exam and a test to see if I would get in. I told my mother, but busy as she was, she did not really take this in.

The next week I returned for a formidable test of arithmetic and strange logic puzzles. Afterwards I was told I had done very well and was up to the standard for the Medical Corps. The medical was easy – the doctor told you to strip off and take deep breaths. He was a local retired man who treated these things as a formality just to get the fee offered. After this we gathered back at the recruiting office. A small Bible was given to each of us and we were told that if we wished to enlist we had to take the oath of allegiance to the Queen printed in the front. A whole group of us took the oath and were then told we were in the Army. Next the conditions and length of service were explained. I was told that "normal" engagement was "nine and three," which means you serve nine years in the Army and then have three years of reserve service to do. To do less than this – "six and six" – meant that you were barred from some army trades – such as PT! I was told that I would get a letter telling me where and when to report.

I told my mother that evening, and she fainted. I guess that the reality of it hit her then. My father was much more phlegmatic. He believed that the army gave you a good training and a great start to civilian life. I had a letter telling me to report to a Junior Tradesmans Regiment (JTR) at Rhyl in North Wales at the beginning of September. I spent the rest of the summer in a very dreamy state trying not to think too much about September.

It was a crisp sunny day when my father drove me down to Rhyl to start my army life, and my spirits soared as we drove. When we arrived my father and I were shown where I would sleep – in a room with fifteen other boys all arriving the same day. The man who showed us around introduced himself as my Platoon sergeant, and my father and I were impressed with this jolly man in a regimental blazer and tie.

That night after my father had left we were all given a wonderful supper of egg and chips – a real treat. Like most of my companions I cried with homesickness that night, but got a good night's sleep.

The next morning the dream was shattered rudely. At 0630 the jolly platoon sergeant appeared in our room, turned on the lights, and shouted that we were to get out of bed immediately. I didn't know some of the words he used but later they became a big part of my vocabulary too. Those who didn't get out of bed quickly had the experience of being beaten with the sergeant's pace stick, a thirty-inch long folding contrivance to measure paces when marching – but it made a great cudgel too. One or two boys did not get out of bed quickly enough and the sergeant then jumped on the beds with his nailed boots on.

And that was how it began.

Just about all my roommates were from working-class backgrounds. Most of us came from the Midlands and the north of England, a high proportion from areas of high unemployment, especially Merseyside and Tyneside. Some had even been encouraged to join the army by their local police so that the police would not have to take further action against them.

A lot of our training was aimed at producing an almost robotic response; a favorite comment among the instructors was, "You are not paid to think!" There was a lot of bullying and weak people like myself often ended up as the butt of practical jokes. What saved me was the fact that I was a comedian and could often, by mimicking instructors or officers, turn the jokes away from me – and also there were boys in my platoon who were much weaker than me.

To cause discomfort or outright terror to one or another young soldier made the day for most instructors. A few were kind and taught properly, but the majority were men who had been sent to the training camp to get them out of their own regiments; many had drinking problems or were plain bullies. The only reason they ended up in the JTR was because their own units had wanted rid of them. They had absolute power over the boys and could inflict all kinds of sadistic punishments on us often for very minor offences. Sometimes you saw boys running on the spot in full kit with rifles or pounding around the drill square in their best uniforms – just to teach them to obey.

One day our whole platoon was punished for some offence – probably for not getting ready in time for some duty or other. After lunch we were paraded outside our hut and told we should go in and get changed into working dress – in two minutes. When we came outside again we were told to go in and change into football kit and so it went on all afternoon – different dress or outfit every two minutes. At last, as we stood outside gasping in the winter twilight, we were told to go in and get changed into working dress again and be prepared for a locker inspection in half an hour! By this time most boys' lockers were disaster areas, and there were many tears of rage shed as our instructors screamed at us that we would miss supper if we did not have a good kit inspection.

Nearing the end of our junior year we had to undergo gas training. This is to be able to use CS (anti-riot) gas, and also to experience it. We were marched across our training area, and as we passed a hedge a voice shouted, "Gas! Gas!" and someone behind the hedge lobbed a CS gas bomb into our path. We all struggled to get out our respirators and pull them over our faces. I had an extra problem: I wore glasses, which meant I had to take off my normal pair and put on my special "respirator" pair.

You were supposed to hold your breath, slip on your respirator, and then blow out as hard as you could, to clear any gas that might have entered as you put your mask on. We'd practiced the drill a hundred times or more, but the real thing is always different. Several boys panicked and sucked in breath as soon as they had the mask on, and then, feeling the effects of the gas, they promptly pulled off their masks, which only made things worse.

Soon several boys were running around coughing and vomiting on the grass as they tried to breathe. Some ran a long way in dire panic before being stopped. I saw one unfortunate lad put his mask on back to front – the straps were over his face! He had completely panicked and lost all reason. I think the instructors had to grab him before anything worse happened. Not that CS will kill you, but it will give you a very unpleasant time for a while. Of course you sweat like crazy in your panic and the gas sticks to sweaty parts of the skin and causes tremendous irritation to skin not covered by the mask. The scene when the gas had blown away was quite horrific: dozens of boys lying coughing and choking on the grass.

In a video on U.S. military enlistment called "Signing Up: It's Your Choice," a recruiter says, "We don't make 'em join the Army. We just give 'em an option. They want to better themselves, and we try to help them out." But the most important, and least discussed, aspect of a military career is left to a high school guidance counselor, who adds:

They make students believe that they can join the military, that they can go to exotic places. But they don't bring home the reality that you can die very easily, that you can be killed.

Bill Wiser, a friend of mine, experienced this firsthand when he journeyed to the Gulf between December 28, 1990 and January 6, 1991, in an attempt to understand the soldiers' outlook better. He was taken aback by the disbelief, among enlisted soldiers, that they would ever have to see live action. On the trip over, a security guard at New York's John F. Kennedy International Airport said to him, "I don't want war, but I may be called. And I don't want to fight, but I'll do what I have to

do. I'm in the reserves. I made a choice, and sometimes you choose to destroy yourself."

In Saudi Arabia, Bill talked with a serviceman from Georgia:

I don't have anything against serving my country, but this is something more than that. We were told we'd be given thirty days notice if we left the country. We were given two hours. I have two boys in school, and I didn't even get a chance to say good-bye. We were told we'd get home shortly and now it's been five months. My wife tells me she's trying to be Mom *and* Dad...If this thing happens there's going to be a lot of bloodshed, and the closer it gets the more scared I get.

After meeting with many more American soldiers, Bill wrote:

The troops are the least keen to fight. Again and again I was told, "I don't want war. I don't want to see action. I don't want to fight." What do they want? They want to get home to their wives and kids. If there is war, they will fight. Why? They have no love of the country that is hosting them. They have no love of the country they may be liberating. They simply signed a contract, and they will fulfill their obligation to the end.

In January 1991, he received a phone call from the mother of an American serviceman stationed in Saudi Arabia. She was part of a network of parents of Gulf War soldiers, and she told him:

Nobody wants to see war break out. We don't know anyone who wants to fight...The leaders seem to think this is a computer game, but it is our sons and daughters. My son comes from a working-class family, white. He wanted to better himself, to get a good education and a good job. What has happened was part of the risk he took, but none of us thought of the reality of actual conflict. With the end of the Cold War we thought we were looking at peace.

My kid is just a child. Sure, he's twenty years old, but he's still just a grown-up kid. He doesn't know if he's more scared of getting killed than of killing someone. The whole thing is so awful.

Later, Bill heard from the mother of another soldier, one who had seen action during the ground war in Kuwait. Between bouts of tears, she related:

He's having a real struggle with what he is seeing there. He was just sobbing – beyond sobbing. Young kid from Iraq. They had killed him. "He looked just like my brother Brad when he was younger. There he lay, dead, with his eyes staring up at us." It makes me sick. How will he handle this when he gets home?

Manuel Carvalho learned these same lessons the hard way. He was sent to Vietnam early in the war to set up the air base at Da Nang.

What eighteen-year-old knows what democracy is? Or what picking up a gun and shooting someone is like? They don't tell you that side. They tell you that they're "looking for a few good men," that you're going to get a nice uniform, that you'll look pretty for the girls. But they don't tell you the reality.

And the reality comes in many ways. I enlisted in a time when there was no war, in 1963. I thought it was a good time: I'd get it done, get it over with, and put it in the past. And then I found myself in boot camp, and then in Okinawa, and suddenly I'm on the shores of Vietnam right after this Gulf of Tonkin business.

Well, to make a long story short, I wound up in a firefight in 1965 and got my legs smashed up by some rounds, and lay in a hospital back in St. Albans, in Queens, New York, reading about the war I just left, and I couldn't believe it. The stories, the misinformation! Only then did I start to believe that I was being conned. I served, I got shot up, I got purple hearts, bronze stars, got paraded by Mayor John Lindsey in New York City and hey, hey, hey.

I had a run-in with one of my neighbors after getting out of the military. Met this commander who was a recruiter, a naval commander, right in Brooklyn, and he asked me, "Well, how was your time?"

I said, "Pretty tough, but I hear you are a recruiter. And you know what? We are on opposite sides."

"What?"

"We're on opposite sides. I'll do anything I can to stop anybody from signing their life away."

Because that's exactly what it was, signing your life away. But you don't realize it. You're eighteen, you've got the world by the tail, and you have no worries…

"My recruiters were in the newspapers, in the magazines, in the films, on the radio, billboards, everywhere," explains Jay Wenk, a World War II veteran and member of Veterans for Peace from Woodstock, New York.

Growing up in that milieu, I was, like all my friends, all of them, dying to get into it. Couldn't wait until I was old enough to get into it. I started to change my mind as I got closer to the sound of the guns. But nevertheless, I was there, I went. I was very scared, I was very shy, I was very naïve, I was very insecure. That's a hell of a place to be carrying a rifle. I carried that rifle in World War II in Germany and chased the Germans from one end of Czechoslovakia to the other. When I went into the Army I was not quite eighteen. When I went to war, I went to defend democracy…but I did not know what democracy was.

Before I went into the Army in 1941, Pearl Harbor was attacked. Every billboard along the roads, every publication like *Life* magazine, was loaded, absolutely loaded with propaganda. All you have to do is dig up any publication from those times, those years, and you'll see what I mean. There were pictures of dead Americans lying on beaches in the South Pacific. There were signs that had been put up on the beach next to these bodies. The signs said, "Kill the bastards," referring to the Japanese, of course, or whoever it was.

Sure, we can understand the anger and the heartbreak that people feel when their friends have been killed. But we also have to understand the profound and crazily insidious propaganda effected by that kind of thing when it's repeated over and over and over and over and over a thousand times a day, day after day, year after year: kill the bastards. Immediately when hostilities break out, we are the good guys, whatever country it is, and they are the bad guys, and they deserve to die. We learn this...

We usually don't know that we're being propagandized when we are. I hope that, whoever has to face the choice of going into the military in order to get an education, realizes that they may get an education different from what they are looking for...

How do you know when you're being conned? When somebody is offering you something that looks very nice at a very low price. What I like to do is to go back inside myself and be in touch with that part of me that wants something for nothing. That's one way of knowing that you're being conned.

Like Jay, Pete Murphy also fought in World War II. He's dying of cancer now. Speaking of his combat days, he keeps his words simple, raw:

I joined the Marine Corps when I was twenty years old and had a wife and two children. I went to Parris Island, and they lied to me all the way.

I hated everybody. Then one day I killed a soldier. He was Japanese. His wallet came out of his jacket, and I opened it up and I looked at it. And there was a picture of him, his wife, and his three children. And I said, "What the hell am I doing here?" Here's a guy that's never done a thing to me, and yet I had to kill him because his boss said "go to war" and my boss said "go to war." And they lie and they lie and they are still lying to us. And they are ruining this country. Your son goes, and do you think they care if he comes back? No they don't. I've seen it with my own eyes.

Ben Chitty served with the Navy in the Gulf of Tonkin during Vietnam. Upon his discharge, however, he became an outspoken member of Vietnam Veterans Against the War. Ben contributes an important observation about veterans and their children in time of war:

You go to the movies, you watch television, you read books, war heroism is looked up to, war is presented as a glorious experience. I was raised on stories of Cowboys and Indians, and the Cowboys always wore white hats and they always won, and of course you would want to be one of those good guys, one of those heroes, one of those people that won. And yet a veteran – particularly a veteran of combat – knows that it's not such a good thing.

One way that you can tell if somebody is really a veteran is to ask him, "Do you really want your son to do what you did?" Most veterans would say no. They don't have any choice about themselves anymore, but they do have choices for their children.

Paul Pappas, a World War II veteran of the U.S. Marines, recently reflected on boot camp indoctrination and the "fruits" of his training:

In spite of not wanting to get killed, I chose to serve in the most dangerous branch of the forces, the Marine Corps. Movies depicting the South Pacific had emphasized only the glory of fighting the wicked Japanese. During boot camp the base commander spoke to us: "Everything in your training has a purpose." How true his words were I did not grasp until later.

Not being an athletic person, I found the physical training strenuous, and it was not easy to keep up. But what was most devastating was the psychological manipulation aimed at making me what I was not and did not want to be – a killer.

I didn't realize this fully until right before the invasion of Okinawa. They did everything they could to make us hate. The caste system of rank was rigidly enforced, and we were constantly kept on edge with spot inspections and forced marches in the middle of the night – whatever could be done to aggravate and anger us. It got so bad that someone blew up the NCO's privy.

The control over us was relentless; there was no room to breathe or relax. All we did was to react – inwardly, not outwardly. Anger, frustration, and hatred built up inside of me; tenderness and compassion were destroyed, and we became beasts, killing machines. There came a point where they had to draw the line, because somebody was going to get killed and it wasn't going to be the enemy.

We were pawns, and we were controlled by powers over which we had no control. The world was going to hell and we could do nothing about it. But it was years before I was able to sort out my feelings or even speak about them.

To veterans like Paul, it's no secret that in times of war governments and armies work hand in hand to engage the general public in their campaigns, heightening the sense of obligation and duty to a feverish pitch. Clearly, something far beyond

human influence is at work when countless families fall in line and enthusiastically support the sacrifice of sons and daughters to "the cause." Defying sanity and reason, these forces compel society, along with its soldiers, to plunge forward into a mission that can only be described as suicidal.

In December 1937, Japanese soldiers killed an estimated 260,000 to 350,000 people in the newly established capital of the Republic of China. Chinese-American author Iris Chang reports that during the atrocity, which was to become known as the Rape of Nanking:

> Between 20,000 and 80,000 Chinese women were raped – and many soldiers went beyond rape to disembowel women, slice off their breasts, nail them alive to walls. So brutal were the Japanese in Nanking that even the Nazis in the city were shocked. John Rabe, a German businessman who led the local Nazi party, joined other foreigners in working tirelessly to save the innocent from slaughter by creating a safety zone where some 250,000 civilians found shelter.

Atrocities on this scale do not happen overnight. As Chang explains:

> In trying to understand the actions of the Japanese, we must begin with a little history. To prepare for what it viewed as an inevitable war with China, Japan had spent decades training its men. The molding of young men to serve in the Japanese military began early: in the 1930s, toy shops became virtual shrines to war, selling arsenals of toy soldiers, tanks, rifles, antiaircraft guns, bugles, and howitzers. Japanese schools operated like miniature military units. Indeed, some of the teachers were military officers, who lectured students on their duty to help Japan fulfill its divine destiny of conquering

Asia and being able to stand up to the world's nations as a people second to none. They taught young boys how to handle wooden models of guns, and older boys how to handle real ones. Textbooks became vehicles for military propaganda. Teachers also instilled in boys hatred and contempt for the Chinese people, preparing them psychologically for a future invasion of the Chinese mainland. One historian tells the story of a squeamish Japanese schoolboy in the 1930s who burst into tears when told to dissect a frog. His teacher slammed his knuckles against the boy's head and yelled, "Why are you crying about one lousy frog? When you grow up you'll have to kill one hundred, two hundred chinks!"[5]

4 The Change is Forever

One guy I'll never forget would come to work with a canvas bag full of kittens. He would go out back and fill up a pail of water and put those kittens in the water, and then he'd come around and open the window next to his desk so he could hear them cry. That was the type of thing that was going on inside of him.

John Risser, U.S. Navy veteran

In an age rife with images of holocausts in Poland and Germany, Bosnia, Rwanda, East Timor, Tibet, and many other places in the world – an age in which we measure the carnage on television by the number of violent deaths per hour – it may be harder than ever to imagine people so affected by traumatic experiences that they can no longer function normally. But we delude ourselves if we think such people do not exist.

Take Charlie, a Vietnam veteran I've known for over three years now. Not long ago, he left New York state and headed for California, back to his roots. He had some important decisions to make, he told me, and he wanted to try to regain his health, which had deteriorated from smoking, obesity and diabetes. His greatest battle, however, has to do with the defining experience of his life: even after thirty long years, Vietnam won't go away.

In his letters, Charlie pours out the anger and hurt still boiling inside. It's not a steady barrage, but it's there, and it comes out in sudden fits and spurts. He went to Vietnam to sacrifice his life for his country. The first American to see him home called him a murderer. She was a flight attendant, and at first he thought he had heard wrong. He hadn't. Those words touched off an ongoing journey to try to understand just what it was that had happened to his life.

Before he left for the West Coast, Charlie went through a month-long program for Vietnam veterans at the Veterans Administration (VA) hospital in Montrose, New York.[1] Near the end of it, he invited me down for the Watch Fire barbecue. I didn't quite understand what it was all about at first, but I learned something that night.

The Franklin Delano Roosevelt VA Hospital sprawls along the eastern banks of the Hudson River in the town of Montrose. The picnic grounds are at the water's edge and are very beautiful. They would have been more so drenched in sunshine, but the day was windy and gray with a fine drizzle. When I arrived, there wasn't a soul around, and I began to wonder if I'd come on the wrong date. I noticed a massive pile of wood in the center of the broad peninsula – huge logs and tree roots, pallets, broken picnic tables, doors and framing – about twenty feet high and forty feet in diameter.

Suddenly, the group began to arrive and, after a few false starts, the barbecue was in full swing. I felt warm acceptance from this group of combat veterans, but it was clear that deep inside they were carrying something outside of my experience. It was written on their faces and bodies: the enormous burdens they had shouldered for so many years, the struggle to cope with horror and despair.

One veteran, lank and wasted with strain, wore a jacket from his Marine battalion in Vietnam. It was covered from top to bottom with medals. He said he had jumped out of helicopters into hot LZs forty-five times; when he was told he would get the "big medal" if he made it to fifty, he had "told them to go to hell." He talked as though it were only a small thing, but it was obvious that those jumps haunt him still.

I was having such a good time mingling among the crowd and enjoying the food that I almost forgot about the fire. The

Watch Fire is surrounded by tradition. Legend has it that George Washington used to light watch fires all up and down the Hudson, to call back lost patrols and to help them navigate in the dark. For the Vietnam veterans of FDR, its meaning has become symbolic. The fires are lighted to "call back" the hosts of war dead: the multitude of lives cut short by a senseless and stupid war, the thousands of MIAs and POWs still unaccounted for. I had thought the lighting of the Watch Fire would be a solemn, ceremonious ritual, but it happened when one of the vets looked up from his hamburger and called out: "Whaddya think? Should we light it up?"

I don't think I have ever seen a fire that huge and that hot. When it was fully ablaze you couldn't stand within fifty feet of it. One vet told me it would burn for three days. As it burned, smaller fires began to appear on the opposite shore in the gathering darkness. There must have been twenty or so in Montrose alone. I looked around at the faces of the men once the novelty of the lighting was over. This was the solemn time, and it was clear that the Watch Fire was not merely superficial, but an experience that stirred deep.

One veteran whispered to me that the guy with the Marine jacket used to burn whole villages with a flame-thrower, sometimes killing women and children and animals. Some bowed their heads. Others covered their faces with their hands. Still others looked deep into the flames, lost in a place and time from which they are unable to escape.

DESPITE FALSE IMAGES of cool, collected Rambos and James Bond-types braving mortal danger to kill in the line of duty, it is clear that there is something about the *real* experience of war that has the power to cripple the human psyche permanently. In 1990, a landmark study mandated by the U.S. Congress concluded that over half of all Vietnam combat vets suffered from significant psychological and emotional problems. These problems, collectively termed post-traumatic stress disorder or PTSD, include depression, alienation, isolation, anxiety, rage-reactions, intrusive thoughts, problems with intimacy, psychic numbing, emotional constriction, and self-defeating and self-deceiving behavior. The study also revealed that ninety percent of these afflicted men and women (8000 of them nurses) had never been near a VA hospital or sought therapeutic help of any kind.[2]

PTSD has also been linked to physical illness. Compared to Vietnam veterans who saw little combat, those with PTSD are much more likely to have circulatory, digestive, musculoskeletal, respiratory, and infectious problems, up to twenty years after military service.[3]

PTSD is not unique to the men who served in Vietnam, though specific factors of that war – the lack of clear objectives and an ill-defined enemy, for examples – contributed to an especially acute epidemic among its veterans. It is common among people of all walks of life who have been through events considered to be beyond the bounds of "normal" human experience. It is an aftereffect of all war, independent of time or place, and it certainly isn't new among veterans, although its name has changed from war to war. In the American Civil War, it was called "soldier's heart"; in World War I, it was "shell-shock"; in World War II, "combat neurosis."

According to a 1998 two-part documentary made by Testimony Films for the BBC titled *Veterans: The Last Survivors of the Great War,* shell-shock cases among British troops officially numbered 28,000 in early 1918. "The real figure was perhaps three times that, and in the worst cases, men suffered a complete physical and mental breakdown."

In his documentary video *Beyond Vietnam,* Steve Bentley (former chair of the Vietnam Veterans of America Post Traumatic Stress Disorder and Substance Abuse Committee) details a government study of returning World War II veterans that revealed such uncomfortable facts about the long-term effects of war that its findings were suppressed. The study revealed that at least 300,000 of the 800,000 U.S. ground soldiers who saw combat in World War II were psychiatrically discharged.

In 1947, the U.S. military commissioned the famous film director John Houston to make a movie titled *Let There Be Light.* They asked him to go into the VA hospitals with his cameras. He did such a good job of capturing the kind of inner turmoil and horror that these men were going through – the survivor guilt, the suicidal thoughts, the flashbacks, the nightmares, the depression, the anxiety, the fear – that the military suppressed the film until 1982. They wanted to silence the evidence which showed that an abnormal response to an abnormal situation is normal behavior. In other words, if some things don't make you crazy, you aren't very sane to begin with. One of those things is war...Given enough time and enough exposure to combat, all normal human beings will eventually break down.

CLAUDE EATHERLY COULD easily have died a hero. Instead, his life stands as a stark reminder of the invisible wounds war can inflict, particularly on a man who attempts to expunge from his conscience the ugly shards of mass murder.

At 7:15 in the morning of August 6, 1945, Claude Eatherly, commander of the lead reconnaissance plane *Straight Flush,* was seated in the cockpit. His mission was to reach the target of Hiroshima, Japan, ascertain the weather situation and the possibility of receiving enemy fire from air or ground, and radio back. Scattered stratocumulus clouds floated toward the city of Hiroshima at an altitude of 12,000 to 15,000 feet, but the bridge on the outskirts of the city was clear. The weather seemed ideal. At 7:30 a.m., according to the official history of the United States Airforce, Claude Eatherly radioed back a coded weather report which sealed the city's fate and forever etched the word "Hiroshima" into the consciousness of human history. Uncoded, the report said "Bomb primary [target]."[4]

At 8:15 a.m. Japanese time, the B-29 bomber *Enola Gay* released its terrifying cargo over the city. Among the messages scrawled on the bomb was one that read: "Greetings to the Emperor from the men of the *Indianapolis.*"* Captain Robert Lewis, the aircraft commander, saw the massive, blinding flash of the explosion and called out: "My God, look at that son-of-a-bitch go!" In that instant, 80,000 people were killed, and

*Eleven days earlier, on July 26, the American cruiser *Indianapolis* had arrived at Tinian Island with the atomic bomb on board. On July 29, the ship was torpedoed at midnight between Tinian and Guam. Approximately 350 of the crew of 1,196 were killed in the explosion or drowned. More than 800 men floundered in the sea, only to be driven mad by salt water ingestion or eaten by sharks in the following days. In all, 883 men died in the *Indianapolis* disaster, the greatest loss at sea in the history of the United States Navy, and the last major warship to be lost at sea in the Second World War (Gilbert, 709).

more than 35,000 injured. Jacob Beser, one of the crewmen, commented, "It's pretty terrific. What a relief it worked."[5]

News of the *Enola Gay*'s success arrived as President Truman was eating lunch aboard the cruiser *Augusta,* on his way back from the Potsdam Conference. Grasping the hand of the map room officer who brought him the radiogram, Truman said, "Captain Graham, this is the greatest thing in history." The President tapped his glass with a spoon and announced the news to his dining companions. While they cheered, Truman spread word about this incredible bomb to others on the ship. "We won the gamble," he said, smiling broadly.[6]

On the ground beneath the blast, a frantic Japanese MP raved and ranted at American POWs being held at Hiroshima Castle: "Look what you have done! One bomb! One bomb! Look there: that blue light is women burning. It is babies burning. Is it wonderful to see the babies burning?"[7]

A group of twenty soldiers huddled in a clump of bushes inside the city, motionless, dying. One called out for something to drink. "Their faces were wholly burned; their eye sockets were hollow, the fluid from their melted eyes had run down their cheeks." Perhaps they were anti-aircraft personnel, and had turned their faces upward as the bomb went off. Their mouths were swollen wounds which could not stretch enough to drink from a cup. So a kindly priest found a large piece of grass, drew out the stem to make a straw, and gave the men water.[8]

The force of the explosion was unlike anything ever seen. Birds shriveled in midair. People died in nightmarish ways:

their skin peeled off, their brains, eyes, and intestines burst, or they burned to cinders while still standing. A Jesuit priest reported: "In the Hakushima district, naked, burned cadavers are particularly numerous. Frightfully injured forms beckon to us and then collapse."[9] Victims who could still walk made their way to one of the seven rivers in the city to slake their consuming thirst. Thousands, burned and irradiated, with skin sloughing off, died there, choking the rivers with corpses.

When Major Eatherly returned to the island base of Tinian to await his demobilization, he spoke to no one for days on end. The members of his bomber group acquired instant worldwide notoriety, and they basked in the glow of Hiroshima. Eatherly's moroseness wasn't taken seriously. "Battle fatigue," they called it. After all, he had succumbed to a nervous breakdown two years earlier, following thirteen months of continuous patrol duty over the South Pacific.

The rest of the crew went on as usual, joking and cursing, slapping each other's backs, and reliving the high points of the big day. But Eatherly's battle fatigue wouldn't go away.

In the months to follow, crew members played to an adoring press. Colonel Paul Tibbets, the "decent, patriotic, and efficient" pilot, said he would do it all over again for the cause of democracy. In an article for the *New Yorker,* he reflected, "People keep asking me these days whether I wasn't shaken by the importance of our atomic bomb work. Myself, I found that I was just as anxious as ever to finish up and get back to that steak dinner and 'Terry and the Pirates.'"[10]

Eatherly remained the black sheep, the sole member of the bomber group who refused to be enshrined as a hero of war.

Twelve years later, he was arrested and tried in Abilene, Texas, for breaking into two U.S. post offices at night. His defense revealed an interesting summary: apparently, Eatherly felt responsible for the deaths of the Japanese at Hiroshima, and "wanted punishment." In the years following the war, he had also run guns for a Central American revolution, forged checks, held up grocery stores, and been in and out of mental hospitals. According to his psychiatrist, he was so tormented by dreams of Hiroshima that he had begun to involve himself in criminal activity, hoping to bring about punishment on himself. The chasm between the voice of conscience and the accolades of society was so great that he was driven to desperation in an attempt to bridge the gap.

In 1959, Claude Eatherly's illegal exploits mysteriously came to an end, and he began to speak out against the horrors of nuclear war. Until this point he had been considered "safe" enough for voluntary hospitalization; now, a Texas jury suddenly found him to be deranged, and he was committed indefinitely, and against his will, to a mental hospital. Obviously concerned that Eatherly's outspokenness might disturb America's postwar contentment, the federal government collaborated with the Eatherly family to ensure that his ravings would be confined to the corridors of an institution.

Vienna philosopher Günther Anders maintained an active correspondence with the captive Eatherly, and later published most of the letters in his chronicle, *Burning Conscience*. In the book's preface, Bertrand Russell writes:

The case of Claude Eatherly is not only one of appalling and prolonged injustice to an individual, but is also symbolic of the suicidal madness of our time. No unbiased person, after reading Eatherly's letters, can honestly doubt his sanity, and I find it very difficult to believe that the doctors who pronounced him insane were persuaded of the accuracy of their own testimony. He has been punished solely because he repented of his comparatively innocent participation in a wanton act of mass murder. The steps that he took to awaken men's consciences to our present insanity were, perhaps, not always the wisest that could have been taken, but they were actuated by motives which deserve the admiration of all who are capable of feelings of humanity. The world was prepared to honour him for his part in the massacre, but, when he repented, it turned against him, seeing in his act of repentance its own condemnation...[11]

Robert Oppenheimer, director of the "Manhattan Project," which created the atomic bomb, is said to have reflected on the prophetic words of the Bhagavad Gita as he watched the first test explosion in the New Mexico desert on July 16, 1945: "If the radiance of a thousand suns were to burst into the sky, that would be like the splendor of the Mighty One...I am become death, destroyer of worlds."[12] But in the aftermath of Hiroshima, which good American really felt he had?

Popular sentiment celebrated with Churchill and Truman: the bomb had supposedly saved a million American lives and brought a quick end to the war. "Up to this moment," Churchill later recalled, "we had shaped our ideas towards an assault upon the homeland of Japan by terrific air bombing and by the invasion of very large armies." The British prime minister waxed eloquent as he envisioned an end to the nightmare in "one or two violent shocks." And he went so far as to suggest

that the Japanese would be grateful for the sparing of further calamity, should these "shocks" be successful.[13]

Truman privately balked at the thought of having to kill "all those kids" in a second bombing, this time of Nagasaki, but he was effervescent in public. He spoke of guarding the new secret to ensure that its power would be an "overwhelming influence toward world peace." And he thanked God that it had come "to us, instead of to our enemies; and we pray that he may guide us to use it in his ways and for his purposes." He confided further to his diary that it was "certainly a good thing for the world that Hitler's crowd or Stalin's did not discover this atomic bomb. It seems to be the most terrible thing ever discovered, but it can be made the most useful."[14]

Against the backdrop of such feverish self-justification and manufactured Providence, Claude Eatherly's voice was unacceptable and had to be silenced. After escaping from his involuntary incarceration in a Waco, Texas, VA hospital he wrote to Anders: "This country is much like many other countries. It is nearly impossible to go against the military."[15] In another letter, he reflected:

> One has only one life, and if the experiences of my life can be used for the benefit of the human race, then that is the way it will be used; not for money nor fame, but because of the responsibility I owe everyone. In that way I will receive a great benefit and feel relief of my guilt. If I were to receive the money for any other purpose, it would only remind me of the 30 pieces of silver Judas Iscariot received for his betrayal. Although it has always seemed to me that the *real* culprit responsible for the judicial murder of Christ was the High Priest, Caiaphas – the representative of the pious and the respect-

able, the "conventional good people" of all ages including our own. These people, while not blameworthy in the same sense as Judas, are yet guilty in a more subtle but also more profound sense than he. This is the reason why I have been having such difficulty in getting society to recognize the fact of my guilt, which I have long since realized. The truth is that society simply *cannot* accept the fact of my guilt without at the same time recognizing its own far deeper guilt. But it is, of course, highly desirable that society should recognize this, which is why *my* and *our* story is of such vital importance. Now I accept the fact that I am unlikely to bring about that recognition by getting into scrapes with the law, that I have been doing in my determination to shatter the "hero image" of me, by which society has sought to perpetuate its own complacency.[16]

Newsweek magazine attempted to squash rumors about the "mad bomber" with evenhanded rationalizations: Eatherly was simply suicidal, and Hiroshima in itself could not adequately explain his behavior; further therapy would surely uncover the real reasons for his disposition.[17] While the establishment explained him away, Eatherly was busy writing, mailing off letters and essays on war and peace to supporters in Europe and Asia. Some of his statements appeared in Tokyo newspapers in 1957, and in 1959 the leaders of Yuwa Kai, the Japanese Fellowship of Reconciliation, wrote back:

We believe that you were acting either under the orders of your superiors, which you could not disobey, or under impulses of war psychology into which men and women are driven in wartime in any country and engage in horrible, inhuman actions without realizing what consequences they entail...We regard you as a victim of war in much the same way as those who were injured in the war and are praying for your complete recovery.[18]

In further letters to Anders, Eatherly poured out his feelings of guilt. He was unable to forget his act and now felt that war was "wild and inhuman." And while he found no understanding among his fellow Americans, he received comfort from the most unexpected quarter of all: a group of thirty young Japanese women, all victims of Hiroshima:

Dear Sir, July 24, 1959

We, the undersigned girls of Hiroshima, send you our warm greetings. We are all girls who escaped death fortunately but received injuries in our faces, limbs, and/or bodies from the atomic bomb that was dropped on Hiroshima City in the last war. We have scars or traces of the injury in our faces and limbs, and we do wish that that horrible thing called "war" shall never happen again either for us or for anybody living in this world. Now, we heard recently that you have been tormented by a sense of guilt after the Hiroshima incident and that because of it, you have been hospitalized for mental treatment.

This letter comes to you to convey our sincere sympathy with you and to assure you that we now do not harbor any sense of enmity to you personally. You were perhaps ordered to do what you did, or thought it would help people by ending the war. But you know that bombs do not end wars on this earth. We have been treated with great kindness by the Christian people (Quakers) in America. We have learned to feel towards you a fellow-feeling, thinking that you are also a victim of war like us.

We wish that you will recover soon completely and decide to join those people who are engaged in the good work to abolish this barbarous thing called "war" with the spirit of brotherhood.[19]

A letter from the Japanese XY Society brought him similar encouragement:

We are continuing to pray for your complete recovery and entry into a new life. This letter we hope will reach you on or around the historic Hiroshima Day. It is a day dedicated to the cause of reconciliation among the nations, a day with a memory, which hereafter must serve to deliver mankind from mutual suspicion, enmity, and war, but to live instead in mutual trust and with brotherly love.[20]

"Whatever may happen in the future," Eatherly wrote, looking back, "I know that I have learned...things which will remain forever convictions of my heart and mind. Life, even the hardest life, is the most beautiful, wonderful, and miraculous treasure in the world...Cruelty, hatred, violence, and injustice never can and never will be able to create a mental, moral, and material millennium. The only road to it is the all-giving creative love, trust, and brotherhood, not only preached but consistently practiced."[21]

On July 7, 1978, the *New York Times* reported:

Claude Robert Eatherly, who, as a young Army Air Corps pilot, picked a hole through the clouds over Japan on the morning of August 6, 1945 and radioed the B-29 *Enola Gay* to drop its atomic bomb on Hiroshima, died of cancer last Saturday in Houston. He was 57 years old...

In 1962, he was [one] of four persons at a demonstration in New York, including Pablo Casals, given "Hiroshima Awards" for "outstanding contributions to world peace."...After his funeral, Mr. Eatherly's brother James, of Midland, Tex., told reporters: "I can remember him waking up night after night. He said his brain was on fire. He said he could feel those people burning."[22]

DENNIS FLAHERTY JOINED the British Army as a fifteen-year-old and spent his prime years on its payroll. Today, as President of Veterans for Peace UK, his time is dedicated to the cause of British veterans and to educating the public about the futility of war. A short, stocky man of sixty-nine, he lives in Beddau, a sullen onetime mining town in Wales, not far from Cardiff. The determined lines of his face seem somehow borrowed from the landscape.

Though he trained as a technician and served with the Royal Electrical and Mechanical Engineers (REME) Corps, his thirst for combat carried him to Aden and Borneo, Northern Ireland and Vietnam. In the end, it completely drained him. Scene One:

We were sent to what was then the British colony of Aden, on the Gulf of Arabia. At the time, Britain was planning to hand the territory back to the local inhabitants, but there were two factions within the local population. They couldn't make up their minds who was going to take over the country when we pulled out, and they had this ideology that the best way to get the British to leave quickly was to kill them.

Our tasks as an armored regiment included patrolling the roads, escorting infantry on foot patrols, and guarding the border. I was a sergeant. We were sent to Habylain where we had an airfield and a camp containing one company of the Irish Guards, a mortar unit from the 45 Royal Marine Commandos, a REME workshop, an armored regiment, and a detachment from the Engineers to look after the airfield. This is where I first came face to face with combat. The Special Forces warned us that a major assault was coming in on us: "Dizzies" (our name for them), seven hundred strong.

Our fire base was a plateau, with the airfield on top of it. And we had positions down the left-hand edge of the airfield. From there the ground dropped off into a very deep dried river bed. This wadi was several hundred yards across and maybe fifty or sixty feet deep. We lived in tents surrounded by sandbags, with earth piled up against those. We had gun positions and fortified enclosures where the armored cars would drive in and the only thing you could see was the gun turret. We had some of the latest equipment, including an electronic radar set that would track a shell back to its source, so that if somebody fired at us, they could only get off a couple of mortar rounds before we could pinpoint them and hit back. Everything was camouflaged with nets. And on the hilltops surrounding us, the Marines had fire bases: 105 PAC Howitzers, artillery pieces, mortars and this sort of thing.

The news came about these approaching Dizzies, and we were given our orders: The armored regiment would mount up at midday, very visibly in broad daylight, and most of the infantry would pile into the armored trucks. We would all head down the road away from the airfield, go out into the desert, and set up camp. And then the minute it got dark, we would travel straight back into camp with the lights out, no noise, nothing. And that's exactly what we did.

Well, of course the Dizzies thought we'd evacuated, but we'd actually gone out into the desert, circled around, and come back again and gone back into our armored positions and waited.

They came at us on foot. They had two Russian-made six-wheeled armored trucks, mounted with 20mm Degtyarev cannon. The rest were on foot carrying Kalashnikovs, mortars, grenades, and stuff like that. The Dizzies came down the wadi, and they got all the way to the wire before we were allowed to open fire. We saw them coming nearer and nearer, but we were constantly reminded, "Don't open fire, let them get right up close." You know, the old Western ideology, wait until you see the whites of their eyes...

We had spent almost the entire afternoon putting tracer rounds into the belts of machine gun ammunition, and the effects were incredible. Usually, every fifth round is a tracer. Well, we had belts that were almost all tracer. So we could see exactly where the point of impact was. The routine was, you fired a burst of ammunition from the machine gun mounted alongside the main armament. And when you saw the machine gun bullets hit the target, you pulled the second trigger, and the main armament fired, which on an armored car was a 75mm short-barreled Howitzer. We were using canister; sort of like a 75mm shotgun shell, only containing ball bearings, not lead shot. Now you're going to fire that when your enemy is less than thirty yards away, and there's seven hundred of them packed into a wadi trying to scale the walls. And you're going to open fire. And the people surrounding you are going to drop 81mm mortar shells on them and 105 Howitzer shells of high explosive.

That was my first taste of real war, and it was like a butcher shop. The first guys had actually got into the barbed wire when the first rounds were fired. Machine gun fire just cut them to pieces, and the canister rounds just dismembered them. Bits of body flew in all directions.

And when the sun came up the next day, we were left to do the cleanup job, walking around with sticks, picking up bits of human bodies and trying to keep the vultures away. There weren't enough bits to say how many you'd killed, but we estimated at least 170. A torso here and a leg over there, an arm and bits of human intestines. And you're in the desert, so the place is literally black with flies. It looks like the wadi has changed color, from sandstone and amber and tan to black, and it's all moving. It's alive. You're walking around with a hanky over your mouth, and you can't get the smell out of your nostrils. We put everything in forty-four gallon drums, dumped in oil, and then burned the lot.

After Aden came Northern Ireland, a chance for Dennis to enhance still further his reputation as a "complete nutter." Not surprisingly, when he ended up working a routine job in an ammunition depot, the boredom soon got to him. Offered a

chance to transfer into the Royal Australian Army, he jumped at the opportunity. He left the British Army and joined the Australian 2nd Cavalry Regiment as a sergeant major – only to find on arrival that the regiment was sitting on the decks of an old aircraft carrier, the HMAS *Melbourne*, which had been converted into an assault ship, destined for Vietnam. Scene Two:

The 2nd Cavalry Regiment was on its way to Vung Tau, Phuoc Twi province, South Vietnam. We were Number 3 Task Force of the first Australian Army unit to arrive in Nam. This was 1967. When we arrived, our American allies told us there were no VC in Phuoc Twi province. They'd cleared them out; we were safe to land, they said. The beach was like a picture postcard: beautiful sand and coconut palms, and off on the left were three hills, about a couple hundred meters high. It looked gorgeous.

The U.S. Marine Corps had lent us a ship with landing craft. The Cavalry Regiment was to be the first ashore, to provide armored cover while the rest of the equipment came ashore. So they decided that the Light Aid Detachment (LAD) would go in front. We had about fifty-odd assault troops in carriers, plus four assault vehicles, and we went ashore in the first wave of landing craft. When we hit the beach everything seemed to be okay. We got the armored vehicles out. The sand was fairly soft, so we moved them up towards the tree line. As we headed up the beach, there was incoming artillery fire. And we discovered that somebody had some old French 75mm guns up in those hills, and they were firing at us. Now we had a choice: we could either get back on the landing craft and go back to the ship, or we could wait for somebody else to deal with them, or…what?

The only fuel and ammunition we had on shore at the time was what we had in the vehicles. Being the senior NCO on the beach – the cavalry officer who was supposed to be in charge of the squadron was already on a landing craft on his way back to the ship – I decided that the best form of defense was to attack. So I got the blokes together and I said, "Come on, we'll take these guys out."

We charged in on the enemy position on these three hills, and we wheedled them out of there. It was a pretty heavy fire fight for about two and a half hours. And we actually chased them off the place, and we destroyed the enemy guns. Yours truly was told to appear before the divisional commander, the Task Force commander. I was told I was an idiot, that I was crazy, that I shouldn't have done it, and that they were sending me back to Australia. So I'd actually been in Vietnam for about ten days all told, and then I was on a plane going back to Australia, thinking I'm going to get court-martialed and all sorts of things.

When I got back they sent me to a place up in Queensland. And I discovered I had just volunteered to join the Australian Special Forces. Nobody asked me. And there I was. In my own tiny mind at the time, I thought I'd made it. I was a senior NCO, and now I was in the Special Forces. I was one of *the* elite.

They put me through a school of jungle warfare. By this time the Australians had my British Army records, and they'd worked out that I'd been a nutter in the British Army, that I was a first-class technician, and that I'd volunteered for every active service tour of duty that was available to me. So they sent me back to Vietnam as an advisor, and I and another guy were sent to educate the Montagnards. Our job was to train these indigenous highlanders in the use of modern weapon systems so that they could act as a buffer between us and the VC and serve as our scouts. So each Montagnard village in the highlands became an intelligence gathering site.

The idea was that two of us members of the Special Forces would train the Montagnards and at the same time use them to do reconnaissance. We were sent out as trail watchers. We would cross the border into Laos and Cambodia. We sat for weeks on end, watching the Ho Chi Minh trail and sending information back about who was moving up and down.

We used to capture VC – lone messengers, couriers that were running the trails – to get information. If you wounded a VC courier on the trail, obviously you couldn't take a prisoner; there were only two of you. So you'd either hand him over to the Montagnards, or you'd kill him yourself. If you caught a bloke, you beat the crap out of him to get information, and then when you'd finished beating the crap out of him you killed him and you buried him and you moved on. You were constantly moving from one hidden position to another, watching, taking notes and relaying information.

Well, by this time I really thought I was it. I was so good at doing this hiding in the jungle and relaying and getting intelligence back and dealing with the enemy, setting up ambushes. I had a hell of a reputation and I was known as "Mad Paddy" to everybody. I was regarded as completely off my trolley. A very, very dangerous animal indeed. They moved me down into the Mekong Delta. They said, "Well, all we need you to do is look after this village." It was full of Montagnards who had been moved into this "secure hamlet," having been forced out of their villages because of combat or because the area had been turned into a free-fire zone.

So now I was in charge of a sixteen-man team with a whole village under our control. The Montagnards in our hamlet knew how to use modern weapons, and we had Claymores and razor wire – all the accoutrements of modern warfare. But we were still human beings. We used to take in all the orphaned Montagnard kids.

We were right on the edge of Phuoc Twi province, and there wasn't much VC activity. But every now and again there would be a fairly large intrusion. We received intelligence from the U.S. Army's 199th Light Infantry that about four hundred North Vietnamese regulars were coming into our territory. So I left eight men to look after the hamlet, and the other eight of us went out to catch these guys on the hop. We set up a really big ambush. It was the middle of the wet season, stinking hot, pissing down with rain. And no one showed up. We were about fifteen klicks (kilometers) away from the hamlet. And then, as we watched, there was an arc-light – a squadron of B52 bombers, at 40,000 feet, dropping 1000-pound bombs. You never hear them coming; first thing you know is when the first one goes off. And we thought, that's awful close to our position. Well, the enemy hadn't showed up so we started back.

When we got back to the hamlet, the only thing that was left of it was craters forty foot across and twenty feet deep – all eight of my men, my team, everybody in the village had been annihilated, there was nothing left. Our entire village, everything in it, had ceased to exist. There were eighty-four children in that village, and there was nothing left at all. Not a damn thing!

I went completely demented, because we knew it couldn't be a mistake; nobody can make a mistake like that. We were nowhere near a free-fire zone, we were nowhere where an arc-light was required. The VC that we'd been told about didn't exist. And what I'd worked out in my tiny brain was that it was deliberate. It seemed to me that we were meant to have stayed in the hamlet, and the arc-light was supposed to wipe us out, to a man. But later, when I thought about it some more, I realized that it probably wasn't us they were after at all. Our neighbours were the 199th Light Infantry Brigade, U.S. Army, run by a bloke called Hackworth. He was a colonel who had said publicly that America couldn't win the war in Nam, that none of us could. I think today that they made a mistake, they bombed the wrong hamlet; they were meant to bomb his, which was the next hamlet along.

Next thing we knew, they were sending the helicopters in. They picked us up, the eight survivors, flew us to Saigon, put us on a plane, and sent us home. The war was over. Just like that. One minute you were in the jungle, you'd lost all your friends and all the kids you'd looked after and all the good things you thought you were doing, and then all of a sudden you were back home in Australia.

They gave us three weeks R&R, and then they sent us back to Nam again. Only they sent us to a different place, gave us different jobs. You couldn't ask questions, nobody would tell you anything. I thought, this is all wrong. We're not fighting a war, we're not meant to win. This is all completely crazy, we're doing all the wrong things for all the wrong reasons.

And I went back in with another team. It was shitty! We were in a very bad area, we were getting hit almost daily, hourly. You know, it was like you're fighting on the run. Hit by the VC, you pull back, you're hit by the VC, you pull back. We were getting chopped to bits. It was unbelievable. And then the Australian government passed a decision saying that the Anzac forces would pull out. The war was over for us altogether then. They sent the helicopters in for us and pulled us out – and we were in the middle of a bloody firefight! It was all over. Very strange feeling, it was all over again. We all got into the Hueys, heading home. We would go home and no more killing and no more living like animals.

We were up there at about 9000 feet, on the way to Saigon, from where we'd be flown home. And one of the guys who had been one of our eight survivors – he was my best mate – he got up and stepped out of the door. It just happened so fast...

I went home, got married, and had two children. I used to wake up in the middle of the night screaming, because all I could think of was this guy stepping out of the door of the helicopter. And I got real weird, real aggressive. I used to knock my kids about. I drank like a fish. Everybody steered clear of me. I began taking out my aggression on my own troops. A new officer, fresh out of the academy, tried pulling rank on me, and I beat him within an inch of his life. So the army decided that was it: I was psychologically unbalanced and I either faced court martial for assaulting an officer or I could resign. So I resigned.

My wife had also come to the conclusion that I was crazy, and she didn't want me in the house. I couldn't stand to have people anywhere near me. You only had to look boss-eyed at me and I'd take you down. I used to drink a lot so I could sleep and so I could get on with living. But I thought, I can't take this. I can't stand to be around my own kids, I can't stand to be around my wife. I couldn't stand to be around anyone. If they would come too close to me I would get really upset, angry, really aggressive, dangerous.

So I got a job with a company in Papua New Guinea putting together mining shovels out in the jungle. I thought, I can handle the jungle, and my wife won't have to see me, and I won't be near my kids, I won't hurt them. All I have to do is send them money. I worked there for two and a half years. One morning, two brown envelopes arrived. One was from a court saying they had an injunction which forbade me to set foot on my own property or come within five hundred yards of my wife. The second one contained a notice saying that my wife was divorcing me and that the court had given her everything I owned. They'd sequestered my bank account, given her my house, my car, my land – everything I owned except the clothes on my back, my toolbox, and $5.00.

They'd taken me to the cleaners. I'd had it. I wanted to kill. I wanted to go home and murder my wife. In fact, I would have killed anybody. It was devastating beyond human belief. And then I had a telegram the very next day, to say both of my parents had been involved in a car crash back in the U.K. My mother had ended up in a wheelchair and my father was partially disabled. I thought, Jesus Christ al-bloody-mighty, the bastard up there is shitting on me like there's no forever.

I came home to England. I had some money in an account my wife hadn't known about, and I used it to buy a house for my parents. I moved them down to South Wales, but I couldn't stand having them around, so I went to work for a construction company in the mining game. I used to work sixteen to twenty hours a day; that way, I came home exhausted and could sleep. Then I'd get up and go back at it. I made a lot of money, but I couldn't stand people. If somebody complained to me about anything, they got wiped out immediately.

I got a contract to go to Saudi Arabia with the mining company, building the main highway there. When I came back from that job, I returned to my parents' home. I started to pass out for no apparent reason, in the middle of the day or whenever. I'd be working, and I'd just go face down. So they sent me to the local hospital to find out what was wrong with me, and they couldn't find anything wrong.

I met a nurse and started going out with her. She was great. We got married, and we bought a house and settled down. My son was born. But I still kept having these nightmares every now and again. I couldn't understand it. My wife would say that I used to get very violent in my sleep and would wake up thrashing about and kicking and biting. And I thought, "Oh no, I'm not going back down this road again, I can't live like this. Life has got to come to an end, I can't take anymore."

I spoke to a friend of mine, and I said to him, "Look, I'm coming apart at the seams. I'm going back to the way I was when I came back from the Nam. I can't live like this anymore and I'm going to end it all." He said, "I want you to talk to a friend of mine." And he took me to see this friend of his who was running a workshop. She was a veteran. Inside of about fifteen minutes, I was huddled in the corner, curled in the fetal position, crying my eyes out. And I thought, this is it, I'm cracking up, I've had it. I just came apart. And when I finished crying, I discovered this lady was sitting right in front of me, holding me in her arms, and she was telling me that I was not to blame; that I didn't do this because I wanted to do it; I did it because it was what I was educated to do. And it suddenly dawned on me that everything I'd done for thirty years was all wrong! I'd been educated ever since I was a child to do what other people wanted me to do, to be the thing that other people wanted me to be. And the information I'd received since I was knee-high to a grasshopper was about revenge, about killing. It was all about hating ...

War is unbelievable. It's a crime against humanity. But what you go through afterwards, when you've survived, is even worse. You become less than human. You become terrified to talk to other people, because you're actually afraid of what you might do to them. We're actually trying to protect other people, which is why soldiers never talk to you when they come back from the war. They don't want to hurt you, they don't want you to know what it's really like; it's too sickening, too painful, too dehumanizing. So they don't talk about it. They think they're protecting you, but the reality is, they need to tell you what it's really like: the horror of it – the absolute unmitigating misery and suffering of it – so that you never go and do it.

Dennis's days of flashbacks and blackouts are over. He has worked through his pain and is able to speak openly about his past, using his experiences to help others confront their own. He's one of the lucky ones. Many thousands of other Vietnam veterans with PTSD continue to suffer deeply from a war that ended more than a quarter-century ago.

Joseph Hughes is one of them. When I first interviewed him at a shelter for the homeless in New York City, I was surprised by his relaxed demeanor. His eyes were alert and attentive, and the hint of a smile played on his lips. He looked every bit the average, middle-aged Irish-Catholic, and I thought to myself, "They've got this one pegged wrong. This guy is well-adjusted."

Joseph seemed almost happy to tell his story, though he spoke in staccato phrases, clipping off articles and pronouns like some kind of radio commentator. His memory was incredible, and he never hesitated as he spoke. But I soon realized that behind the mask was a deeply hurting man whose heart and mind had been irreparably damaged by Vietnam, where he served with the U.S. Army's 124th Signal Battalion, 4th Infantry Division, stationed at Pleiku Base Camp. His cynicism,

distrust, and professed absence of humanity were almost frightening, yet at times I could sense something of a tiny spark within – a flicker of soul in a man still crying desperately after thirty long years. Joseph told me how it all began:

We were told to avoid civilian casualties and stuff like that. That was the official line. Then you get in-country and you listen to the guys who've been through it already, who've got six months, or maybe a year. And you don't trust any of them. You don't trust any Vietnamese. I don't care if you're in town, in a whorehouse, out in the field, on a convoy, or later on in the PX (post exchange). And I was still kind of a boy scout, you know, Peace Corps volunteer, did some work in Honduras in '64. I still had that crazy idea in my head that you had to differentiate. But I got straightened out on that real quick, by the senior guys.

One time when I was new with the company I rode shotgun on a water truck to the other side of the air force base, to get 5,000 gallons of water for the base camp showers. And I caught a little kid. My buddy was in the bushes with a Madam K, you know, one of the truck-stop hookers. I'm a new guy. I have my rifle loaded up, and I'm outside the truck. And I catch this ten or eleven-year-old kid putting a hand grenade in the gas tank with scotch tape around the handle. Gas eats away the tape. I couldn't believe it, I was brand new. I grabbed him. He had the grenade in one hand, the pin down, and I got the kid. These women were screaming and yelling, and I dropped him. I panicked. Locked and loaded, and I shot at the kid. But I missed him on purpose. I could have killed him easy, but I didn't. And that's always bothered me. I've told this to a few VA psychiatrists, and they say, "Well, it's good you didn't, because you'd feel guilty." But I feel guilty that I didn't kill the kid, because he was a combatant. Just as willing and able to kill me or any other GI. And so that's tough shit for him. I should have killed that little bastard. Bothers me to this day.

We had a commando patrol about a month later. Walking down this road, and three guys step out of the tree line. And everybody opened up. It's a miracle we didn't shoot each other, because the way we were staggered, bullets flying past my ear, I'm shooting past this other guy ahead of me. One of the slopes went down. He rolled into a ditch, and I remember running up to him. He had a head wound, brain tissue coming out, and this medic comes up from out of nowhere and starts putting an IV into his arm. And a tall, lanky infantryman kicks the gook right in the face. He says, "Get the fuck out of here! You save that for us." Any kind of medical care you get, you don't waste on the enemy. So he kicked the slope in the face, and shit went flying, blood and brain tissue. That's when I started to get hard. 'Cause the E7 was right, I mean, fuck the gook. He's the enemy. I don't care if it's man, woman, or child. They're the enemy. That's it. They die. And no thinking about it. You can't feel sorry for the enemy in a time of war.

Vietnam was, I think the fancy word is "superlative." Any emotion was maxed out, totally maximum. If you were scared, it was the most scared you'll ever be. If you were bored, it was the most intensely boring time you've ever gone through. If you hated somebody, it was the most intense hatred you'd ever felt. So it was a lot of boredom, punctuated by extreme excitement, extreme rushes. Some of the guys were very special soldiers. I mean, they volunteered again and again. They'd get high on war. And I experienced some of that. I can't put myself in their class; I don't have that honor. But I experienced some of that same high.

It's still there. I always thought the war was righteous, and I still do. And it really galls me, it infuriates me that we lost 58,000 men and maimed 300,000 for life, and it was for nothing. Because the big shots screwed up bad...

I wasn't home a couple of weeks when I stepped off the deep end with the drinking. I wasn't home two weeks when I looked like my Uncle Joe, a heavy-duty alky... And then the girls I had been dating before Nam:

Gabbi Wiser, she says, "Who the hell are you? I don't even know you anymore." Mary Brennan said the same thing after a couple dates. Well, fuck them. Plenty of women around. Who needs them? And things have been downhill ever since.

I've been homeless most of the time. Very depressed. My life's been a mess. Put my wife and my mother through nothing but torture. It's almost impossible for me to make a decision, stay with anything. 'Cause the Army doesn't believe in that. That's your camouflage, the haircuts and uniforms. And you hide in there. Guys who tried to maintain their individuality, they got the shit knocked out of them. It's all group think.

Been mostly unemployed. Twenty-eight years. My anniversary's coming up, September 19, when I came back. That's a lot of years, and I've been unemployed more than half of that time. A lot of drinking, a lot of pot smoking. You know, last six, seven years been all drinking, out of work all the time. Divorced. Like I said, I have this big conflict. I would do Nam again. I don't know. It sounds crazy. Makes me sound like heroic. It's just unfinished business.

When I got back from the Nam I just didn't belong. Odd man out. I'm still standing on the outside, like I tell my psychiatrist over here. Twenty-five years and I feel like I'm on the outside of civilian society, and I'm starving to death, and everybody's getting invited to a sit-down Christmas dinner with all the trimmings. And I'm starving. And I can see it. I can see everybody sitting down. I go around and get in line. When they open the door I can smell the aromas. And as soon as I go in, the civilians just close the door and say, "Not you. Step aside. Throw this bum out."

So fuck the civilian world. Fuck it. I like to borrow something from what the Marines say: no such thing as an ex-Marine. And it's true. No such thing as an ex-serviceman. No such thing. So fuck the civilian world. Excuse my language, but fuck it.

WALTER WELLS GREW UP in New York City's South Bronx and served as a helicopter door gunner in Vietnam. He has never really returned:

Within thirty days of coming home from Vietnam I was in prison. I was carrying a gun. I mean, how can you take a person out of combat and expect him to act differently, without having any type of resocialization program? They had no deprogramming when we came home. They just said, "Washington – New York – Good-bye." From that day on I've spent close to twenty years in the penitentiary, in and out of prison. I've been to Attica. I've been to Sing-Sing. I've been to Napanoch, I've been to Collins Correctional Facility, Elmira, Wallkill, Downstate, Peekskill, Clinton. The big ones. I came home in 1971; 1990 was the last time I was in prison.

I cry about it sometimes, wish I never came back. It's hard for me to talk about my feelings, about what I've been through, and living here in the shelter, losing my family because of drugs, not really getting any help... You know, nobody's really reaching out to help or to do anything for us. I mean, we can go to the VA; we can sit and talk about what we've seen in Vietnam, but nobody really understands: how you can take a young man out of the ghetto, set him into a situation like that, and then bring him back home where people spit at him and call him a baby-killer and shit like that. Then you have Desert Storm, and everybody's a fucking hero. War is no good for nobody, man.

I have a lot of pain inside. I like walking in the rain by myself, in a storm. I like to be alone. And that's when things are real dangerous with me, when I'm alone, because I'm with my own mind. I dream. I pray a lot, and that helps; God's been good to me lately. But I'm still in a lot of fucking pain.

WITHOUT A DOUBT, PTSD is most acute in veterans who have seen or participated in the killing of defenseless people, es-

pecially women and children. Though such killing is a war crime by definition, it happened over and over at the hands of Americans in Vietnam and elsewhere.

On the morning of March 16, 1968, a group of American GIs called Charlie Company entered the sleeping hamlet of My Lai in Quang Ngai Province on the coast of Central Vietnam and, under orders from their leader, Lt. William Calley, proceeded to murder more than four hundred old men, women, and children. In the course of four hours, they torched homes and pagodas, slaughtered water buffaloes and chickens, raped women and children at gunpoint, and filled irrigation ditches with more than four hundred brutally mutilated bodies.

Though Lt. Calley was eventually brought to trial and found guilty of premeditated murder (he was sentenced to lifelong imprisonment and hard labor, but was paroled after serving less than four years), incidents like My Lai remain white-hot with controversy because the guilt lies somewhere between the soldiers involved and the factors which drove the war: the reality of woman and child combatants, the VC's use of villages for protection and hiding, the booby traps, mines, and snipers, the pressure for "body counts" and "kills," and the dehumanization of the Vietnamese people as "gooks" and "dinks." In the final analysis, lives were ruined on all sides – the villagers of My Lai, the young men pushed to the brink and scarred to the depths of their souls, and a little bit of each one of us. As a GI mother later cried, "I gave them a good boy, and they made him a murderer."[23]

Many Vietnam veterans will confirm that My Lai was not an isolated incident but happened again and again. That this is probably true does nothing to mitigate its horror; rather, it suggests that what happened to the men of Charlie Company in the wake of the atrocity can be extended to thousands of others.

In a harrowing account of the massacre and its aftereffects, British authors Michael Bilton and Kevin Sim relate an interview with My Lai GI Varnado Simpson. It is a glimpse into a private and unimaginable hell.

At the end of April 1982, a 34-year-old black "Vietnam era" veteran was admitted to a Veterans Administration hospital in Jackson, Mississippi. His name was Varnado Simpson. He was of average height, around 140 pounds, well dressed, and well groomed. Although he possessed above-average intelligence and a good vocabulary, it was observed that he was nervous as he spoke, smoking continuously and making wringing movements with his hands. He always sat with his back to the wall and would never allow anyone to get behind him.

According to their notes, Varnado Simpson told doctors at the hospital that he had entered military service in 1967 and had been posted the following year to Vietnam. There his company had attacked the village of "Milai" and were "apparently" ordered "not to leave anyone alive." He himself had killed women and children in the village and described "other very traumatic events occurring during this action."

When he returned home from the war, Simpson started work in a bank. But in 1969, with the revelations of what had happened at My Lai, followed by the trial of Lt. Calley, he had left his job. People in the street called him "child-killer" and "baby-killer," and he felt that customers were withdrawing their money from the bank because he worked there. Since then, he had become reclusive, extremely fearful, and "somewhat paranoid."

Nightmares had become so frequent that he was afraid to go to sleep. The people he had killed in Vietnam, he said, were not really dead and were going to come back and kill him.

In 1977, Varnado Simpson's 10-year-old son was playing in the front yard of his home on the northern edge of Jackson. Teenagers across the road began arguing and one of them pulled a gun. A wild shot hit his little boy in the head.

> I was in the house. And I came out and picked him up. But he was already dead…he was dying. He died in my arms. And when I looked at him, his face was like the same face of the child that I had killed. And I said: this is the punishment for me killing the people that I killed.

Doctors could do very little for Varnado Simpson. They described his condition as post-traumatic stress disorder, "chronic and very severe." In the year he remained in the hospital they had the greatest difficulty in trying to get him to talk about his experiences. Any discussion or activation of his memories from Vietnam created such extreme discomfort that he simply could not tolerate it. "Mr. Simpson has been one of the most uncomfortable people that I have ever seen," his doctor wrote. His prognosis was considered "very guarded and very poor."

In July 1983, with his doctors' agreement, Varnado Simpson went to live by himself in a small house on the same street where his son had died. He barred the windows and put a battery of locks on the doors. In his own house, he felt better able to control his own fate.

Six years later, in the summer of 1989, he still lived alone in the same house. There were more bars on the windows now, and more locks on the doors. The curtains were always drawn. Not much of the bright, hot Mississippi summer filtered indoors.

Inside, the living room was gloomy. Varnado Simpson's hands were still shaking wildly. He tried in vain to rest them on his legs, but his legs shook too. His whole body shuddered in distress. There was the same thin moustache he wore in the twenty-year-old photograph taken during the war, though his face was rounder, almost swollen, and there was a puffiness under the eyes as if he had been crying, or was about to cry. He sat in an old armchair, his head a little to one side, with a look of utter resignation on his face and his back to the wall. And he began to talk about what happened in Vietnam.

"That day in My Lai, I was personally responsible for killing about 25 people. Personally. Men, women. From shooting them, to cutting their throats, scalping them, to...cutting off their hands and cutting out their tongues. I did it."

Why did he do all that? Why did he kill them and do that?

"I just went. My mind just went. And I wasn't the only one that did it. A lot of other people did it. I just killed. Once I started, the...the training, the whole programming part of killing, it just came out."

But your training didn't tell you to scalp people or cut ears off.

"No. But a lot of people were doing it. So I just followed suit. I just lost all sense of direction, of purpose. I just started killing any kinda way I could kill. It just came. I didn't know I had it in me.

"But like I say, after I killed the child, my whole mind just went. It just went. And once you start, it's very easy to keep on. Once you start. The hardest – the part that's hard is to kill, but once you kill, that becomes easier, to kill the next person and the next one and the next one. Because I had no feelings or no emotions or no nothing. No direction. I just killed. It can happen to anyone. Because, see, I wasn't the only one that did it. Hung 'em, you know – all type of ways. Any type of way you could kill someone, that's what they did. And it can happen."

"It can happen." It did happen. One hundred and five GIs from Charlie Company went into the village of My Lai. Almost all of them are still alive, living quietly all over America. Each of them must have their own way of dealing with the things they have seen and the deeds they have committed. But it is in Varnado Simpson's sunless room that the poison of the memory of My Lai seems to survive in its most concentrated form.

"I can't remember – you know. I can't remember everything. I don't want to remember," he says. But the truth is that he never forgets. Remembering has become a compulsion. Suddenly, without warning, he jumps up and pulls a large brown book out of a cupboard. A scrapbook of My Lai. Twenty-year-old photographs and news clippings are neatly preserved in an old photo album.

Why has he kept it?

"This is my life," he answers without hesitation. "This is my past. This is my present, and this is my future. And I keep it to remind me. But it's always there. This is my life. This is everything. This is the way I am. This is what made me this way.

"I have an image of it in my mind every night, every day. I have nightmares. I constantly have nightmares of the children or someone. I can see the people. I can go somewhere and see a face that reminds me of the people that I killed. I can see that vividly, just like it's happened today, right now."

Varnado Simpson remains caught in a trap sprung twenty years ago in Vietnam. He wants to be put out of his misery. He has attempted suicide three times and does not know if he will still be here "the next time you come around." On a table, there are dozens of bottles containing the pills which he takes throughout the day and night. One jar is labeled simply: "For Pain." Nothing prescribed by his physicians seems to have much effect or even relevance to his case. He is stretched out somewhere between life and death, scared of both, convinced that on Judgment Day his will be a hopeless case. For there is little doubt that Varnado Simpson sees himself as a man who has already been damned.

"How can you forgive?" he asks. "You know, I can't forgive myself for the things I did. How can I forget that – or forgive? There's a part of me that's kind and gentle. There's another part of me that's evil and destructive. There's more destructiveness in my mind than goodness. There's more wanting to kill or to hurt than to love or to care. I don't let anyone get close to me. The loving feeling and the caring feeling is not there.

"That was caused by My Lai, the war. My feelings and the way I feel and the way my life is.

"Yes, I'm ashamed, I'm sorry, I'm guilty. But I did it, you know. What else could I tell you? It happened. It can happen if you go to war. Those are the type of things that will happen and can happen to anyone..."[24]

In late 1997, just a few months shy of the thirtieth anniversary of the My Lai massacre, Varnado Simpson took his life.

ALL TOO OFTEN, the anxiety and self-destructive behavior of veterans suffering severe PTSD ends in suicide. Though controversy swirls around the actual numbers – ranging from as low as 10,000 to as high as 120,000 – the Vietnam War continues to exact a heavy toll back home. And as widow Deborah Cook writes, even one suicide – and the awful wake of pain and anguish it leaves behind – is one too many.

During my only visit to the Vietnam Memorial in Washington, DC, I was determined not to make this my catharsis but just visit. As I walked slowly and quietly down the walk my eyes caught a letter almost waving to me in the mild breeze.

I picked up the letter and as I read, all of my personal emotion welled up and the sadness, anger, and bitterness was right there, right there at the foot of all those names of the Vietnam dead. The letter was from a woman writing to her lost love. It said something to the effect: "If you had lived, we would have been married, had children, but you didn't come back." I wanted to find her and tell her my story because my love did come back...

Larry and I met when I was fourteen and he was sixteen. High school sweethearts, we became engaged when I was sixteen, just before he left for Vietnam. I wrote a letter to him each day, sometimes more. He was shot in the leg in the twelfth month of his thirteen-month assignment. He came home, and during his convalescent leave we were married. Ten months later our daughter Laura was born while Larry was in the Mediterranean. He left the Marine Corps shortly after his return.

His problems started shortly after the birth of our son Kevin, two years later. I won't go into all of the detail of the hell we lived through. I will tell you that the last year of his life was torture for both of us. Vietnam came back to him in bits and pieces till, in those last days, he was sure people from Vietnam were in the grocery store, at the gas station, or wherever we went. He became paranoid. He attempted suicide perhaps three times before he was successful. I saw it. He shot himself in the head as I walked through the front door (yes, he was under the care of a doctor).

I was a twenty-four-year-old widow with two babies. I would like to tell the woman who left the letter what it is like to be twenty-four years old and see your husband blow his brains out. I would like to tell her what it's like to tell your kids a thousand times what happened but you can't tell them why.

How did Vietnam change my life? It has been almost twenty years since that day in July. I have lived with it every day. I have cried and screamed so many times it feels like that is who I am sometimes. My kids grew up without a dad. My daughter was married last year; we both cried because her dad wasn't there to give her away. She recently gave birth to our grandson. Weeks before he was born, Laura cried and cried because Larry wouldn't see his grandson. God, I hope he can see him – my joy, my respite in this struggle.

My husband's name is not on the Vietnam Wall, but he is as much a fallen soldier as any. Vietnam will live for a long time in the lives of those whose loved ones fell, both there and here.[25]

Admiral Jack Shanahan, of the Center for Defense Information in Washington, DC, said, "In the military we train people to break things, to kill and to destroy. So if you suddenly put these people in a humanitarian area, they don't know how to deal with it... You kind of don't expect them to return to the private sector of civilian society." I thought that a powerful, albeit reserved, statement about the crippling nature of military indoctrination and combat experience, one that explains why veterans like Joseph Hughes have never really "returned" to the civilian world. Lt. Col. Dave Grossman says it most succinctly: "War is an environment that will psychologically debilitate 98 percent of all who participate in it for any length of time."[26]

STEVE BENTLEY HAS WRITTEN extensively about PTSD since the early 1980s. He was a Rome Plow operator in the jungles of Vietnam, clearing the field in areas of heavy enemy contact, booby traps, landmines, or mortar and rocket attacks. "You can't tiptoe through the jungle in a twenty-five-ton bulldozer," he said. "They know where you are."

Steve has tirelessly advocated the cause of fellow Vietnam veterans suffering from PTSD, a work all the more remarkable for his long personal struggle toward wholeness through a tragic family background, drug and alcohol addiction, and a series of hospitalizations after the war. It was clear, when I met Steve at a retreat for Vietnam veterans in October 1997, that he is still working on the "healing end" of his experience (he

said he felt he was meant to be there, in that time and place, with that group of fellow veterans), but his gift for articulating the pain and struggle, the cry from the depths among so many Vietnam vets, has helped others immeasurably. Steve has "been there." He wrote the following in 1992:

Four months ago, in late November, Vietnam veteran Robert Daigneau left the Togus Veterans Administration Hospital after a 28-day stay on the post-traumatic stress treatment unit. He immediately drove to Kentucky where he shot five people to death, including himself.

A week later, 20-year veteran Bruce Allen returned to Limestone after a 2-week stay at the Togus facility. The night he returned he shot himself in the head with a .357 magnum and died. On January 12, Vietnam veteran Dave Garland jumped to his death from an abandoned mill in Old Town. Less than a week later, Vietnam veteran Michael "Mickey" Obrin returned from Togus and shot himself to death in his Portland apartment. On March 17, Vietnam veteran William Harrington returned from one of many visits to Togus and held at least a dozen police officers at bay for an hour while threatening to kill himself with a hunting knife.

One might ask at this point just what the hell it is that the Togus VA facility is doing to these people that allows for such behavior right after discharge. A letter sent to Maine's congressional delegation by wounded Vietnam combat veteran Frank Muchie on March 20 would appear to provide part of the answer. According to Frank, what the hospital is doing to these people is nothing, absolutely nothing. Frank reports that during his 22-day stay there were days at a time when he had nothing to do but worry and wonder and wander the halls...

All the major studies have warned that as the majority of Vietnam veterans enter midlife crisis, dragging along the unfinished business of Vietnam, their problems will intensify...

There are vet centers. However, they are overloaded and understaffed. And while they should be expanding, some Republicans are pushing to phase them out...Denying these problems won't make them go away.

How can it be that a nation has hundreds of billions of dollars to spend destroying human beings and yet not nearly enough to heal them, and what values were we defending in Europe, Asia, and the Gulf if we have money to send children to war, but none to heal men? If those who spill their blood and sear their souls in service to our country can't get justice from our country, then who in God's name can?[27]

I hope that you

can see the true hell I live through every day. My experience has left me a very lonely person. I have no real friends, and I don't think I ever will until I find peace within myself. I hope that some day I will be able to forgive myself.

Doug

I met Doug at a veterans' retreat in Rhinebeck, New York. He was young, only twenty-nine, but his brooding face, far older than its years, was one I could not forget. He sat in the dark, away from the circle of light that brightened the center of the room.

Over the next few days, I watched as he slowly opened up to the members of our group. And I saw him shake, his lips trembling uncontrollably, as he dared to begin to remember. Blurry things, buried deep: shots fired, gasoline, a dead woman, a dead friend. "Did we volunteer for all this shit? Did we volunteer for the hurt, the anger, the shame, the guilt? Would we still have volunteered?..."

In the months since that week in October, Doug and I have kept in touch. Though hesitant to stir his memories to the surface, I was also determined to learn his story. For Doug was a veteran of a hidden war, a modern warrior who knew the jungles of Guatemala and El Salvador and the streets of Panama like the lines on his hand. He had been trained at the School of Americas in Fort Benning, Georgia, as a sniper, a "pathfinder." He had studied the art of killing at Fort Bragg, and had apparently learned well.

I was unprepared for the envelope that appeared early in 1998, containing Doug's story and a very brief note: "I hope the holidays have treated you well. Here is my story. I don't know if you will be able to use it. It was very hard for me to write. I felt that I needed to share my life so that people could understand war better. I also needed to write this down, so even if you don't use it, I want to thank you for asking me to write. Peace, Doug."

The true hell of war doesn't start until you come home. You are a different person altogether. You become dead. What killing does is to slowly eat away at you until you are dead emotionally. That is the only way that I can describe it. To this day, my experience is still with me. I still wake up in the middle of the night…

It's two o'clock in the morning, and I find myself at the top of the stairs drenched in sweat, crying. The nightmares have come back. This is the first time that this has happened in a long time. I wish Abby were here with me. But like most things in my life, she is gone too. I feel so alone in this world we call life…

It all started eighteen years ago, when I was about ten years old. I learned my first lesson about being a hunter. It was how to walk and move quietly through the forest. I learned this lesson from my father, a man whom I admired very much. He taught me different skills that I would later use in my own survival. I wish I could tell him thank-you today.

There are many events in my life that I don't wish to go into here… You would probably be bored to death, because it would take many more pages to tell you how I ended up where I did. I've also purposefully left out exact dates, because most of the missions I took part in are considered never to have happened. I am writing not to bring attention to myself, but to bring healing to my own life…

I served as a sniper in the U.S. Army, in Delta Force, the covert military group responsible for "antiterrorism." They are "the best of the best." My job took me to all the hot spots in the world, but mainly to Central America. There, I was baptized into hell…

My main job was to wreak havoc upon the National Guatemalan Revolutionary Unity, the sworn enemies of the CIA-backed Guatemalan Army. My first kill was a man who was a local leader of the Revolutionary Unity. At the moment I pulled the trigger, my life changed forever. The time leading up to that point and immediately after was filled with pure adrenaline. It was very powerful. My next five kills had the same effect. We had other engagements with the drug armies, but the kills there were justified, because they had a fighting chance. The sniper kills never knew what happened to them. One minute they were talking to their friends or family, and the next minute they were dead.

After the sixth kill, I was sent to Germany for further training. This was at the time of the conflict with Panama...

Throughout my entire time in the military I was separated from regular soldiers. I liked it a lot, but it was also very lonely. I could talk to only one person, and that was my scout, Michael. He was responsible for watching my back, for determining wind direction and speed, and estimating the range of the target.

One thing you need to understand is that we were inserted into position at night. This took anywhere from a couple of hours to several days. We took down our targets from 600 to 1000 yards away...

Snipers are probably one of the worst agents of destruction there are in a conventional war. Three of the people that I shot were with their family at the time they were taken down. Drug lords. At the time, I had to remind myself that they were the scum of the earth, responsible for the drugs on our streets and the horrible conditions of their own countries. They deserved to die.

At the end of my training I was sent back to the "Ranch," Fort Bragg, North Carolina. At this point I was pretty dead emotionally, and looking back, I can see that I died a little bit with every mission.

At Fort Bragg, I was ordered to go on a mission against a group of people I had never dreamed of – our own soldiers. I was assembled along with Michael and four other men whom I never had met before. We were among the few soldiers in the U.S. Army at the time with combat experience, with confirmed sniper kills; we were also the best of the best.

The thinking at the Pentagon was that to get the soldiers stationed in Panama to fight, they had to have very good reasons. We're talking here about soldiers who have never experienced combat before. And the best way to get them riled up was to attack them. When Michael asked what the other four men in our mission were doing, we were told it was none of our business...

You see, American soldiers, especially infantry soldiers, stick together. If one of them gets into a predicament in a bar – I mean a fight – the others don't walk away, they join in. You don't fight one of them, you fight all of them. Their training has taught them to be a team; they depend on each other, and it doesn't matter if it's a barroom brawl or not. They depend on each other to get home. So what better way to get them all worked up than to take pot shots at them? We were told that we would be saving lives by doing this. For weeks leading up to the invasion of Panama, Michael and I took pot shots at soldiers during the night...

During the invasion, we were to provide support fire for Task Force White, a Seal team that would take patrol boats out of Balboa and secure the Paitilla Airfield. There were a lot of casualties that day. I had never seen so much death and destruction before.

As we were moving back into Balboa, we were attacked by a small group of the PDF and supporters. As we came around a street corner they were lying in ambush for us, and Michael was shot in the left arm by a woman who was maybe twenty or twenty-one years old. Seeing him fall to the ground, wounded, unleashed within me a rage of anger that I had never before experienced. I went crazy. After about thirty minutes we secured the area. I then proceeded to do something that I never envisioned in my life. The woman who shot Michael in the arm was lying in a doorway, screaming for her friends to help. I walked over to her as she was yelling at me, and I slit her throat. It was out-and-out murder.

And it was what we had been taught. It was an attitude of defiance, a way of saying, "You will not fuck with my friends. If you do, I will kill you."

And I did. I killed her over a flesh wound. But it was more than that, emotionally. She had tried to kill the only friend that I ever had.

Several weeks later, my life started flashing before my eyes. That was when the flashbacks started. I realized who I was: nothing but a killer. No better than the people I killed, just luckier. I really started analyzing myself at this point in time, and I knew it would not be very long before either Michael or I would be killed.

Although we can try to hide our emotions, inside we are all human. Humans feel. Michael and I were sent to Fort Benning, Georgia, to the School of Americas, to teach Latin American soldiers sniper skills in a jungle environment. It was considered downtime for us. You work from 7:00 a.m. to 5:00 p.m., eat and sleep and have fun. After about six weeks we were told to get ready to leave; we would be heading out in the morning.

We were never told more than what was considered relevant. And where you were going was never "relevant." It did not matter. I lived within a world of secrets. I never guessed that twenty-four hours later I would be back in Central America and that this would be our last mission together.

As we were getting ready to leave, Michael was reading the newspaper. He showed me an article. The headline said, "The Pentagon denies allegations that American soldiers are involved in the drug wars of Central America." We shared a short laugh.

We were flown into Panama and given our OP orders by a spook (CIA). Our mission was to take out a Salvadoran ex-soldier who was responsible for guarding drug caches waiting to be shipped to the United States. One of the reasons he was a target was that he had killed several Americans in Guatemala and El Salvador. And the other reason was that he was a graduate of the School of Americas.

This mission was very dangerous. We were going to be choppered in to an LZ (landing zone) about twenty klicks away, which I felt would surely alert everyone in the area. On previous missions we had moved in either by foot or by parachute "halo" (high-altitude low-opening) in the night; we'd been picked up by choppers ten to fifteen klicks away. Going in you always had the upper hand, because they never knew you were coming. Going out, you knew it was going to be hot. On this mission, we were going in hot and coming out hot. It gave us a very bad feeling.

We finally decided on two different LZs. If it was too hot at the first LZ, we would move on to the second LZ, which was about forty klicks away from the first. We would have to move about sixty klicks in four days, which, through the jungle, is very strenuous even under the most favorable conditions – and we were being hunted by people who knew the area better than we. We moved into the area with no problem. It was a very quick insertion. We rappelled into a small clearing around 9:00 p.m. We then moved on until daylight and holed up on a ridgeline until the afternoon. Then we moved out, taking the last ten klicks very slowly. We moved into position on another ridge overlooking a small village. Our intended target was to be in this village sometime during the day. As the morning moved on, we could hear more and more movement all around us. We figured "he" had to be coming soon, because of the increasing patrols around us. Around 2:30 p.m. he finally showed up.

It had been a very long forty hours, with little catnaps here and there. For some reason I was very tired. I had no adrenaline running, but rather fear, which had always been there before, but not like on that day. I was done. This was it. When we got out and got home, I was finished. I was not going to do this anymore. And I told this to Michael. He told me I was just scared, and so was he, and what would I do anyway but work at McDonald's? I would feel dead…I told him I was already dead – it couldn't get any worse. I didn't know at the time that it would…

I had been having flashbacks for a while, mostly when I was asleep. Although I couldn't remember the dreams, I would wake up screaming or crying, drenched in my sweat and blood. In my sleep, I had dug at my arms till they bled. I think Michael knew that I was going through this, although he never came out and told me. But he would hint at it. He would say we have done things that we are not proud of, but we had to do what we had to do to survive...

Our target finally arrived, but there was a small problem. It looked like he was with his family. All that went through my head was, I can't kill this man with his family watching. Michael knew this. I was the leader. Everything had to be perfect before I took the shot. I could call this off at any point. The problem was, everything was perfect. There was no reason why I couldn't take the shot.

They started to kick a soccer ball around, and that was when we decided to take the shot. I remember Michael calling the distance and wind off to me. I will never forget that: "550 meters, two klicks to the right." The man's head came into my scope. I pulled the trigger and watched his head explode. I closed my eyes and started crying. I was in my own world. All I could see was that woman lying in the doorway...

I remember Michael telling me to move. "Let's go! Get moving!" I picked up my weapon and put my ruck on my back and we started moving toward the LZ.

The jungle around us came alive with men looking for us. I decided that we would not even go to the primary LZ; we would move on to the secondary LZ. We would be safer that way. If we kept moving, putting distance between us, we would be better off. I have never been so wrong in my life as I was in that decision. We moved on into the night, avoiding the patrols. We stopped around 9:00 p.m. to rest until around 1:00 a.m., when we would move on until daylight and try and hole up for a while. The countdown would then begin, and we would have twenty-four hours to make the LZ and get the hell out.

The area around us was still pretty hot. At dawn we started into a hollow which seemed to be a good spot to hole up. All of a sudden, I had the urge to move my bowels. We settled in about fifteen yards apart so we could watch each other. I moved off to do my business. I had to go; I couldn't wait any longer. All I kept thinking was, "Someone is going to smell this. They are going to find us." When I finished, I couldn't see Michael from where I was. I started to crawl back into my position when I heard gunfire erupting from his position. As I came up on my knees I could see two Indians. They had Michael pinned down. I took the first one out. At that time I don't think they knew we were both there. Michael was able to move back toward me. We were about five feet apart. I lost sight of the other Indian.

We both knew what to do. We had to take him out very quickly, before the others came. It was quiet for a moment or two, and then all of sudden Michael just stood up and started firing at the spot where he thought the Indian was. I started yelling, "Get down! Get down! What are you doing?!" Michael got down. It was too late. I heard the shot and saw the muzzle blast from a small bush. All of sudden I was wet. It was blood and body parts. It was Michael. He was lying in front of me, and he wasn't moving.

The next couple of minutes were just a blur to me. I could hear a man whimpering and breathing very hard. He was hit. I just stood up; I didn't care if I died or not. Nothing mattered anymore. Michael's killer was lying on the ground, out of ammo. I climbed on top of him and just started stabbing him in the chest with my bayonet. I don't know how long this went on...

Michael and I had made a promise a long time ago: no matter what, we would never leave the other to rot somewhere. So I picked up his body and started to head for the LZ. I made it there without any problems. I sat and held him for the next ten to twelve hours. I couldn't move. I couldn't let him go.

Finally I heard the chopper coming. I moved to the edge of the clearing. Once the chopper landed, I picked Michael up and started for it. We were going home.

The chopper and the ride back to Fort Bragg is a blur to me. I was debriefed at the Ranch and I stayed drunk for several days. Something happened to me. I died in the jungle that day. Nothing mattered anymore.

I pulled light duty from that point on. Basically, I reported in during the morning, and at night I was not allowed to leave the Ranch. My enlistment in the Army would be ending in another six months. I knew that I couldn't re-enlist; that was not an option.

I was paid a visit by two spooks (CIA), who told me that I had a job with "the company" if I wanted. At this point, I was a complete wreck. Losing Michael was like losing my brother. And on top of it all, I felt, and still feel, completely responsible for his death. I told my commanding officer I wanted out. I didn't care if it was a dishonorable discharge or not. If he didn't get me out, I would probably be going AWOL. He told me he would find out for me what my "options" were.

A couple of days later he came back and told me the paperwork was started, although I would have to go through several weeks of debriefing. At this time, I didn't care about anything in the world. I just wanted to go home.

I was deemed psychologically fit to be released. I signed a lot of different papers, and I was told that nothing we did ever happened. Point-blank. I was processed out with an empty personnel folder. My DD214 is completely blank. I never existed.

Once out, I realized I couldn't go home. I couldn't face the people I loved. I moved in with some people I knew, but it didn't last very long. I was soon homeless and addicted to coke. It was the only way I could numb my emotions. I stayed this way for approximately three years, in and out of contact with my family. Every time I would get close to sharing with them, I would leave town.

During this time, I was completely whacked-out. Somehow I had managed to get married, but the relationship ended after only four months. I moved in with my parents, but that didn't work either. So I took off one morning and moved into a state forest, living as I had in the jungle. I think I was in flashback twenty-four hours a day for about three months. Close to winter, I decided it was time to go back to society.

I walked out to a small town and called my cousin and my uncle, who was a Vietnam vet. They came and picked me up. We decided that what I needed was a good drink, and we proceeded to get drunk in a local bar. Sometime during the drinking, I walked out of the bar, went to a liquor store, bought a case of beer, and went to get drunk by myself. No one understood me, not even my uncle. I somehow managed to call my mother and told her to come pick me up. I needed to go to a hospital. She took me there, but it was very hard for her to see me hurting so much and unable to talk. I stayed in the hospital for several weeks before going into a private rehab center. Then I moved into a halfway house about four hours away, where I was able to stay clean for about six months.

Then I relapsed and spent the next few months drunk and high. I found myself mixed in again with the wrong people, and I found work running drugs and guns. I was so addicted to coke that I was actually completely worthless except as a user. The only thing that saved me was that I knew how people operated.

It wasn't long before I started robbing the dealers hard. Within a month, I was on the run, living in an abandoned factory during the day and hitting the dealers at night. My adrenaline was back. I was at home. I was safe and secure, always one step ahead of everyone else.

It all ended very quickly.

I went to a farm where I had worked at the halfway house. I stole a pickup truck and headed back toward western Maryland where my family lives. Once back in my hometown, I needed gas for the truck, so I stole it. I broke into several garages looking for gasoline. Somehow, during one of the break-ins, someone saw me and called the police. They found the pickup truck, which had my coat and other belongings in it. I was now on the run from the police, but again I was one step ahead of them. I circled my parents' house, which was surrounded by the police. My father was on the porch. He didn't have a clue as to what was going on. They hadn't seen or heard from me for months.

I proceeded to head for the railroad tracks, hoping to hop a train and be gone. But something happened within me. I knew that I couldn't keep running. I was tired of running. I wanted everything to stop: the dreams, the flashbacks, the memories.

I saw a police car. They were looking for me. So I moved out into the street and lay down spread-eagled in the middle of the road.

My life changed from that point on. I did a year in jail. When I was released, I moved into a halfway house in Baltimore, where I met two men who were working on the street, serving breakfast to homeless people. I started going out with them in the mornings. It was a very healthy experience. The nightmares stopped. We started a small nonprofit group, serving homeless men and women breakfast and going around to different churches, talking to people about homelessness. It was during this time that I met Abby, and we became friends over the next year.

But the peace I had found was slowly leaving me. I found myself needing to be in dangerous situations just so that I could "feel" again. I started going out at night, hoping that someone would try to rob me or kill me, so I could kill them. I needed that release, that challenge.

The people I was working with knew something was wrong with me. They thought it would be good for me to move into the shelter again. I did, and once there, one of the counselors, who had been a chopper pilot in Vietnam, talked me into going to the hospital. I checked into the psych ward of Perry Point, VA...

When I was finally released I went to a halfway house right outside the hospital grounds. I enrolled in a hospital work program. The nightmares were gone, and so was Michael, but only for a little while. Soon I moved in with Abby. But after a couple of months, the anger at myself started showing in different ways. The dreams came back. I was always angry at Abby for some ridiculous reason. I was flipping out again. I could feel it coming on.

Abby and I talked. I checked myself back into the hospital. I came very close to hurting a fellow patient on the psych ward, and ended up in restraints, drugged up in the "quiet room." I was very explosive then... Michael was in my head every moment of the day, and I couldn't share anything with anyone. I couldn't tell anyone my hurt. I slowly started to talk in group session, but I still couldn't talk with Abby.

After my release, Abby and I were married. I knew that she really loved me for who I was – nothing else, just Doug. But my anger was still present, and we rode the waves. It was like being on a roller coaster, although you didn't know how high you were going or how fast you would go down. And it came to an end in June '97. By September, Abby had moved out into a safe place. I did not know where.

Before she left, she had signed us up for a Buddhist retreat in New York State. Part of it was a veterans' retreat, and she also signed me up for that. I was not happy about it, but I knew that if I didn't go, I would lose my family and myself.

I found a lot of healing during this retreat. I was finally able to talk about my experiences, and even though I still held some of them back, I was able to put my feelings into writing.

My thoughts on war and combat are that everyone loses. In some ways, my experience is different from that of Vietnam vets. But in a lot of ways it is the same. One of the ways it is different is that I was never drafted like a lot of them were. I volunteered for everything that I did. But I didn't really volunteer for the emotions and the life afterwards...

In the military, everyone is expendable. It doesn't matter who you are or what you are capable of. They will bring the primal beast out from everyone, and they will train you how to kill another human being. What they don't teach you, though, is how to deal with the emotions afterward. They just spit you out and find someone else to take your place. You do not matter.

The one question that I keep asking myself – and I think more people need to ask it, too – is this: who ever gave our government the right to decide which side is right and which is wrong? Who says that we are on the right side in Croatia? In the Gulf War, were we really there to help the people of Kuwait, or to protect the oil companies of the world?

I think that the only way we are going to find peace is through our children...As one of the most powerful countries on the planet, we need to start finding peace within ourselves. And we need to take drastic measures. We need to ban nuclear and biological and chemical weapons altogether. And the CIA needs to be dissolved completely...

I don't think world peace is something that will be achieved in my lifetime, but we need to start teaching our children that even if people all over the world are different we all have things in common. Instead of teaching about the Vietnam War, why not teach about the people of Vietnam and what a beautiful country they have? Or how about teaching respect for all people, no matter what they believe in?

The Hitlers and Saddams of the world will always be with us as long as we believe that only the rich and powerful are good, that they are the only ones worthy of living. We teach this in different ways. For example, we have nuclear fallout shelters for the leaders of our country but for no one else. To me, that says we are all expendable.

I hope that you can see the true hell I live through every day. My experience has left me a very lonely person. I have no real friends, and I don't think I ever will until I find peace within myself.

I hope that some day I will be able to forgive myself.

6 Worth the Price

It is a difficult decision to make, but we think the price is worth it.

Madeleine Albright, U.S. ambassador to the UN *on the deaths of 500,000 Iraqi children due to economic sanctions after the Gulf War*

One of the most damning aspects of war is that it is never confined to the legions of young men who spill their blood in battle. Invariably it boils over to affect civilians caught between the hammer and the anvil. Even modern technology is powerless to erase this awful fact. "Smart bombs" did nothing to save the 1,200 innocent women and children incinerated in the Al-Amariyah Shelter during the 1991 bombing of Baghdad. And the 100 million landmines currently scattered over 62 nations worldwide have killed more people in times of peace than they did in the wars during which they were deployed.[1]

In fact, technology has become warring mankind's own worst enemy; more, rather than fewer, civilians are killed with each successive generation. One-fifth of those killed in World War I were civilians; in World War II, this figure rose to one-half. In the wars of the past few decades, it has been ninety percent.[2] The modern economic sanction is perhaps even more cruel; it is a form of warfare in which one hundred percent of the casualties are civilian.

Between 1898 and 1910, U.S. forces killed 600,000 Filipinos while seizing the Philippines from Spain.[3] When the Japanese launched a massive attack on the Republic of China's newly established capital in 1937, 300,000 Chinese were massacred and 50,000 women raped and mutilated.[4]

In the war to follow (1939–1945), 6 million Chinese civilians were left dead; Japan lost 2 million. The Soviets suffered the loss of more than 7 million civilians; Poland lost 3 million; 6 million more, all Jews, died in Hitler's extermination camps.[5]

The occupied nations of Western Europe fared not much better. More than 1.5 million Yugoslavs died under German occupation; in Greece, a total of 380,600 civilians died from privation and hunger and execution by occupying forces.[6] France lost 108,000, Belgium 101,000, and the Netherlands 242,000. Great Britain, which was never occupied but suffered prolonged bombing, lost 61,000.[7]

Civilians under the Axis powers were no more fortunate. When the British Royal Air Force raided Dresden on February 13, 1945, with phosphorus and high-explosive bombs, they created a firestorm that killed an estimated 135,000 civilians in one night.[8] Altogether, the Germans calculated 3,600,000 civilians dead.[9] And when more than 100 U.S. bombers raided the city of Tokyo with incendiary bombs on the night of March 9, 1945, they killed more than 124,000 civilians.[10] In Hiroshima, an estimated 100,000 people died the day of the atomic bombing; another 100,000 died soon thereafter from burns, injuries, and radiation.[11]

Of those national or ethnic groups that suffered the loss of a million dead or more, the total dead is in excess of 46 million.[12] Despite attempts to estimate civilian dead, however, the number of those who died in the Second World War will never be known with precision. Millions of men, women, and children were killed without record of their names, or when and how they died.[13]

DURING THE "AMERICAN WAR," Vietnam was poisoned with 18 million gallons of herbicides and devastated by more than 15 million tons of explosives, twice the amount used by the U.S. in all of Europe and Asia in World War II. The war left 3 million dead and millions traumatized, disfigured, handicapped, orphaned, childless, and displaced.[14] Between 1969 and 1975, an additional 2 million neighboring Cambodians died as the result of bombing, starvation, and political chaos.[15]

During the Gulf War, nearly 200,000 people were killed in the invasions of Iraq and Kuwait.[16] The economic embargo that followed Iraq's collapse in early 1991 resulted in a five-fold increase in child mortality, and more than 500,000 children under the age of five have succumbed in the past seven years.[17] Armed conflict has led to unmitigated suffering in Bosnia, Sudan, Somalia, Angola, and the former Soviet republics. According to the Center for Defense Information, in the past decade 2 million children have been killed, and three times as many have been permanently disabled or seriously injured.[18]

It is impossible for the human mind and heart to comprehend such statistics or to take in the magnitude of such suffering. We cannot embrace its impact. Numbers on the scale of millions escape our imagination, leaving us numbed and overloaded. But the legacy of civilian suffering in wartime is a true war story in itself: in the words of Tim O'Brien, it is not moral, it does not instruct, nor does it restrain men from doing the things men have always done.[19] Nevertheless, the stories of civilians who have suffered innocently in times of war can shake us into realization, and change our perceptions and lives.

Lowell LeBlanc, a World War II veteran, served under false identity in the Counterintelligence Corps in England. Stationed

in the rear, he saw very little of the war's horror until the fighting was over. Recently he told me of the experience that changed his attitude toward war forever:

I was drafted into the Army April 1942 at the age of twenty-two, with more or less everyone's blessing and acquiescence. Much like the typical draftee of that time, I was not all that sure or that eager to go. At a medical examination before entering the military, a psychologist asked me how I felt about the Army. I told him, "I'm in, so what's to feel?" Much later I realized that he was right to imply that one should be either for or against it.

Shortly after V-E day, our airfield was used as a landing place to repatriate 1500 people from a concentration camp in Czechoslovakia. I was asked by the owner of a local cafe to go down to the courthouse with him to get thirty stretchers, and when we then drove out to the center of the field, we saw thirty men whom I can only describe as being a few breaths away from death. Some could move, and one or two could talk, but for the rest the only sign of life was an occasional flicker of the eyes.

Death was right there, lying on the ground. They were that close. These men were helpless, and they had no means of altering their situation. It was a turning point for me, as I finally saw with my own eyes the result of man's inhumanity to man.

After what I had experienced, I felt that I could not continue to serve the military. I had initially worked for Boeing, but could no longer be a part of an industry that was hand-in-glove with the military. So I found another job, although the pay was about a third less.

LE LY HAYSLIP WAS A young Vietnamese girl caught in the maelstrom of the "American War." The legacy of her people's suffering, which began long before the first Americans arrived,

plunged many of her compatriots into despair and resignation. But for Le Ly it gradually gave birth to personal transformation and profound insights on the nature of war. She writes:

[We are the ones] who did not fight – but suffered, wept, raged, bled, and died just the same. We all did what we had to do. By mingling our blood and tears on the earth, god has made us brothers and sisters.

If you were an American GI, I ask you to...look into the heart of one you once called enemy. I have witnessed, firsthand, all that you went through. I will try to tell you who your enemy was and why almost everyone in the country you tried to help resented, feared, and misunderstood you. It was not your fault. It could not have been otherwise. Long before you arrived, my country had yielded to the terrible logic of war. What for you was normal – a life of peace and plenty – was for us a hazy dream known only in our legends.

Because we had to appease the allied forces by day and were terrorized by Viet Cong at night, we slept as little as you did. We obeyed both sides and wound up pleasing neither. We were people in the middle. We were what the war was all about...Children and soldiers have always known it to be terrible...[20]

Before he died, Le Ly's father reminded her that she had not been born to hate, and that her main battle in life was to raise her small son.

I decided I should draw the strength of compassion, not the weakness of bitterness, from this most important lesson – from the lessons I had learned from every American that fate or luck or god had sent to be my teacher...Hating people who had wronged me only kept me in their power. Forgiving them and thanking them for the lesson they had taught me, on the other hand, set me free to continue on my way...

Vietnam already had too many people who were ready to die for their beliefs. What it needed was men and women – brothers and sisters – who refused to accept either death or death-dealing as a solution to their problems. If you keep compassion in your heart, I discovered, you never long for death yourself. Death and suffering, not people, become your enemy; and anything that lives is your ally. It was as if, by realizing this, an enormous burden had been lifted from my young shoulders…

My task, I was beginning to see, was to find life in the midst of death and nourish it like a flower – a lonely flower in the graveyard my country had become…

Houses could be rebuilt and damaged dikes repaired – but the loss of our temples and shrines meant the death of our culture itself. It meant that a generation of children would grow up without fathers to teach them about their ancestors or the rituals of worship. Families would lose records of their lineage and with them the umbilicals to the very root of our society – not just old buildings and books, but people who once lived and loved like them. Our ties to our past were being severed, setting us adrift on a sea of borrowed Western materialism, disrespect for the elderly, and selfishness. The war no longer seemed like a fight to see which view would prevail. Instead, it had become a fight to see just how much and how far the Vietnam of my ancestors would be transformed. It was as if I was standing by the cradle of a dying child and speculating with its aunts and uncles on what the doomed baby would have looked like had it grown up. By tugging on their baby so brutally, both parents had wound up killing it. Even worse, the war now attacked Mother Earth – the seedbed of us all. This, to me, was the highest crime – the frenzied suicide of cannibals. How shall one mourn a lifeless planet?…

It was as if life's cycle was no longer birth, growth, and death but only endless dying brought about by endless war. I realized that I, along with so many of my countrymen, had been born into war and that my soul knew nothing else. I tried to imagine people somewhere who knew only peace – what paradise!

...Perhaps such a place was America, although American wives and mothers, too, were losing husbands and sons every day in the evil vortex between heaven and hell that my country had become...[21]

IN THE SUMMER OF 1997, I drove out to Woodstock, New York, to the home of Jay Wenk, a World War II veteran. It didn't take long for us to get down to the business of the interview, and we were soon in his wood-paneled kitchen, discussing the background to the war, pacifism, patriotism, and psychopathology.

Though Jay has seen his share of firefights, they are not among the memories that have left him the most scarred. I was caught off guard when, about halfway through our interview, he shared his "most horrific" war story, one he had never told anyone else in nearly fifty years:

I wanted to share with you my own most horrific story. It's quite different from what the others were. It's at least part of the reason why I feel that young men and women shouldn't go into the services.

Towards the tail end of the war we were almost out of Germany and approaching Czechoslovakia. We captured a town not far from Schweinfurt, where they were making ball bearings and bombs. There was another factory making radios for the Wehrmacht. And there was a small camp full of displaced people or prisoners who worked at the factory. Like many of these labor camps, it was operated by a private firm. My patrol was going out to the eastern edge of town. We took up a spot at the very end of the village.

There was a little cottage where my patrol was domiciled overnight. Every once in a while one of us was supposed to get up and stand guard duty. After it got dark the patrol leader came back. He had gone away some place and came back to our cottage with five or six women. He had given them cigarettes or chocolate or whatever, black-market stuff. And they were brought there to have sex with us.

I got into bed with one of those women and had intercourse with her. It was the first time in my life. The first time in my life, under conditions like that. I don't know how to describe the horror of it. It didn't hit me till later. At the time, you know, you got into this cot and bang, bang, bang, that was it. But later...

Early the next morning the man who was on guard outside rousted us all out of bed. He had seen a German squad up in the woods on a hill right next to us. We went running up the hill and there was a little bit of a firefight. The Germans took off pretty quickly, but there was shooting going on both ways. Nobody was hurt on our side, and I don't think any of the Germans were hit either. In any event, coming back down the hill, going towards the cottage, we saw the faces of these women looking out the window.

We had just been chasing, looking to kill, their husbands, their brothers, their uncles. There they were, looking at us – not with hatred in their faces, but with a kind of curious inquiry of some sort that I can't describe. I still see them looking at me, at us. I felt guilty and embarrassed and tried to justify doing what we had to do. It was a horrible evening, a horrible night, a horrible morning.

I could tell you other kinds of war stories, but that was the most horrific. To use women like that, to go anywhere near that kind of thing, there's a poison there which can infect. Even in so-called peace time. I've never told that story before...

The day before we crossed the Rhine we were in the city of Mainz. I was rummaging around, walking around in the basement of this apartment building. There were chicken-wired cubicles, with people's furniture and God knows what. Locks on the doors. Walking around, I heard a woman cry out. It was nearby. I followed the sound. There was a young woman – I guess she was in her late teens or early twenties – a German woman, and she was holding what looked like a big loaf of pumpernickel. My squad leader was trying to pull it away from her. I said to him, "You can't do that." He gave me a look of disgust and said, "She's German." I said, "She's a civilian, she's a woman, she's hungry. We have plenty of food. We're not starving."

We did have plenty of food. The look of disgust, I won't forget. He turned and left. The next day I was sent to the heavy-weapons platoon, meaning I was loaded down with cans and cans of 50-caliber ammunition. It was painful. The whole incident touched a guilt button. One of my guilt buttons went off, and I remembered that story from "my basement."

ILSE VON KÖLLER GREW UP in a villa on the outskirts of Leipzig. In the early years of World War II she met her husband-to-be Ulrich, and they were married in January 1942. By the time their daughter Martina was born in October, Ulrich was already fighting on the Russian front. Ilse was left with her child and her twin sister in Eisenach, worrying from day to day whether her husband would be killed in action.

She couldn't have known then that before the war was over she would have to fight her own war, and that as a German mother "caught in the middle" she would see plenty of action. She was on the other side of Jay Wenk's war; a German woman looking back at him through the window. Yet she did not look back with "curious inquiry" but with a compassion and understanding not unlike that of Le Ly Hayslip twenty years later. Only a few years before her death in 1995, Ilse remembered:

The war went on and the bombing of the cities got worse and worse. One morning on the radio we heard of a bombardment on Leipzig the night before – 400 American bombers had dropped incendiary, high-explosive, and demolition bombs on the city between 11 p.m. and 7 a.m. One third of the city was destroyed.

My mother lived alone in Leipzig at that time, so I immediately went to find her. I traveled by truck, by train, and also a long time on foot to look for her, to find out if she had survived. When I arrived in the city, a ghostly, still, uncanny atmosphere surrounded me. The smell from the conflagration was terrible, and gray smoke lay heavily over the city. Everywhere there were men from the Survival Service with their long probes, searching for survivors, for a sign of life under the fallen houses. The dead, covered with rags, were lying at the sides of the street. The people had empty faces – sad, confused, hopeless, frightened.

The steeples of the churches were still burning; houses were still smoldering. Some of the streets were completely impassable. The houses, some more than six floors high, were now only a heap of rubble. Under them, people were buried alive. The temperature was -22 degrees C. There was only frozen water, no electricity, no gas.

My mother's apartment was seriously damaged and empty. Where was my dear mother now? A man told me he had seen her running out of the house, raving, confused. I searched desperately from one *Auffangplatz* (a place where survivors could register, get something to drink and eat, and be given a place to sleep) to the next. At last I found my mother, after thirty-six hours of searching.

She was not at all pleased to see me: "Go back, oh, please go back to your child! What will happen to you when the bombers come again tonight?" She didn't want to come with me to Eisenach, so I left her with a friend. It was hard to leave her, but she pleaded with me to go and wouldn't relax until I said good-bye.

Then the bombing got worse in Eisenach, too. I took my Martina and went to my mother-in-law in East Pomerania, in the country. There we were finally able to sleep through the night without running to the cellar when the sirens wailed.

Toward the end of the summer we began to see refugees from East Prussia in kilometer-long lines walking on the highroads – mothers with babies and children and old people carrying all their belongings on their backs, or on top of their prams, or pulling handcarts. The weather turned bad and it rained a lot. The refugees could hardly move forward on the muddy highroads. All these people had lost their homes in the east and were on their way to the west. On top of all their misery, they were often attacked by low-level flyers who shot at the people with their machine-guns. Often there wasn't enough time to run for the ditches, and they were wounded or even killed.

The noise from the Russian cannons grew louder and louder. Day and night the shooting came nearer and the front line rapidly approached. The temperature plummeted below zero. I had to flee to the west with Martina.

A truck drove us mothers with babies and little children a few kilometers westward. There we were herded into a stock-car. We were thirty-five women, lying on straw side by side in the dark. The cracks in the wooden walls afforded little daylight but a lot of draft. Our furniture consisted of an iron stove for heat and for melting snow, a bowl, and a box. A bucket standing in the corner served as a toilet. The heavy sliding door was locked from outside.

Our car was part of a military train carrying the survivors of an air-force fighter wing. They had just returned from an air base in Russia and like us were being forced back to the West. Whenever the train stopped the airmen brought us wood or charcoal for the stove, snow to melt for water, and food from their own rations. We traveled for many days, and because the tracks and bridges were often destroyed, we had to make long detours.

One evening we had stopped on a siding near the city Güstrow, next to two long trains. We were already asleep when the sliding door was pushed open. "Raus! Raus! Come out, come out!" the airmen shouted. The air-raid alarm wailed. We threw our still-sleeping children down into the outstretched arms of the airmen, and they ran swiftly with them into a nearby field. They sat down with our children in the snow. At first I couldn't find Martina, because an officer had her under his big winter coat to keep her warm.

In the sky overhead, American fighters swarmed over the city. First they launched colorful balls, yellow, green, and red, which hung suspended in the air to tell the flyers where to drop their bombs. The sky was strangely beautiful, like a decorated Christmas tree with colorful burning candles. Then German planes flew up to fight them, and at the same time the anti-aircraft guns and artillery started up, sweeping their powerful searchlights across the sky. Desperately the planes tried to escape from being seen and shot. To be illuminated meant great danger; the wings and tails of some planes were already burning. One exploded into burning pieces, which dropped down to the earth. The German airmen next to us watched the fight with binoculars and explained to us women which planes were hit – German or American. I saw seven planes burning, and from one of them, two parachutists sprang out. Where and how would they land?

Then the bombers came – always twelve bombers together. They flew in an arrow, an exact V, like wild geese. One wave after another arrived. They flew very high and then, above the city, they came in very low and dropped their bombs. With great speed they took to the air again and vanished. They dropped more then 10,000 bombs – incendiary bombs, high-explosive bombs – all over the houses, factories, churches, schools, hospitals, ammunition and oil depots.

There was fire everywhere – blue, red, black, and orange. When a spirits depot was hit, we saw the conflagration rise hundreds of meters high.

The whole city was a sea of flames. To see that was more than a human heart could take in, to think of the helpless children, women, and old people. Why did they have to suffer so much misery and distress? Where was God's mercy?

The terror, the chaos, the senselessness took away any human feeling. What remained in people's hearts was dullness, apathy, bitterness, or else thoughts of revenge and hatred. The cry to God, begging for mercy and help, thanking him for sparing us, was missing. We thought only of how we could save ourselves.

Suddenly the air-force officer sitting close to me stood up, gave me my child and his coat, and whispered to another flyer. Then he crawled to the next field, toward some undergrowth. We heard two shots. Then he came back to us and said, "Erledigt (settled)."

I asked him, "What is settled?"

He replied, "The spy is dead; I shot him."

"Why didn't you take him captive first?"

"A spy caught red-handed must be shot immediately," he said.

Behind the undergrowth several flares had been sent up to the bombers in the sky, advising the enemy of the four military trains that stood only a few hundred meters away.

The officer continued with great seriousness in his voice: "You didn't know that two of the trains are hospital trains, full of seriously wounded soldiers with their nurses and doctors, returning from the front. Should they be bombed, too?"

I wasn't able to reply. I didn't know myself what was right to do. The bombing of Güstrow continued till morning. Two-thirds of the city was destroyed, and two thousand deaths were mourned that night.

We traveled on by train. One morning we awoke to find the sliding door ajar, and looking out, we saw that we had been abandoned. Our car had been left on a sidetrack. The military train was gone. After many difficulties

Martina and I managed to find lodging with the burgomaster of Cracow. This family showed us much love, and soon we lived like a real family together.

The Russians conquered more and more German towns. Many institutions and establishments had to be evacuated. We lived on a busy street and saw everything that passed by the house. Once, more than a thousand Russian prisoners were marched down our street on their way to camps further to the West. They were worn and tired, and obviously in need of nourishment, but we were forbidden to give any food or drink to a prisoner. All the same, an old woman from a house directly across the street threw a sandwich out of the window. In no time the men were fighting, and finally some lay on the ground. The long train of prisoners was out of order, and the escort soldiers got furious. They beat the men on the ground cruelly with their rifle-butts. Three of them were unable to get up and were left lying on the side of the street until they were thrown into an open cart at the rear of the train.

Two days later a very different procession passed by, once again heading West. This time, however, those marching were thoroughbred horses: five-hundred beautiful horses, mares and stallions, colts and fillies, all beautifully brushed and certainly well fed. They were traveling from Stolp, a town in East Germany that had a world-famous stud farm. These horses couldn't be used to pull a carriage; they were saddle horses, trained for the races. Their coats shone brightly in the sunshine. During the war they had suffered nothing, no bombing, no hunger, no strafing by low-level flyers. All arrived safely at their destination, where they were confiscated and flown under protection to England…

When the Russian front line stopped at the River Oder, my twin sister Ruth came from Eisenach to fetch us. She was fearful that the Russian army would soon be in Cracow and was quite sure it would be better to be conquered by the Americans than by the brutal Russians. But I couldn't believe that the Russian army would cross the Oder. I had finally found my own lodging, and I wanted to wait until the war was over. Eisenach was hardly habitable and was still being bombed by the Americans. So with a very sad and heavy heart, my sister left us.

It was only weeks before the Russians arrived. The shooting came nearer and nearer, and we looked for a place to hide. Together with the wives of the burgomaster and school principal, we rowed to an island on one of Cracow's many lakes. Martina was two and a half years old. Although the shack had neither windows nor a door, we were grateful for a roof. But it was only days before we were betrayed, and one day twenty Russian soldiers raided the island.

The leader barked and an interpreter translated: "You hid yourselves here because you didn't want to remain in Cracow. You are partisans in the Hitlerpartei, and worst of all, you have high functions in it. I am commanded to shoot you... You!" he pointed at me. "Your name, and where is your husband?"

"Ilse von Köller, refugee from East Pomerania. My husband is in the war in Russia," I replied.

"Yes, you spoke the truth. I'm informed about you all."

To be German was bad enough, but the aristocratic "von" was even worse. In 1918 the nobility had been shot in Russia...

We had to stand in a line against the shack. Five soldiers stood in front of us with their rifles in hand. The leader counted and the rifles were pointed at us. My only thought was, "What will become of my child Martina when I am dead? Abandoned, left alone in the hands of these brutal soldiers?" I saw her in my thoughts, unhappy, desperately crying, "Mutti, Mutti!" Then, still in my thoughts, she was going slowly to the shore, and she fell into the lake. I saw her little body lying on the bottom of the lake. That would be more than terrible. I took her in my arms; she had to be killed with me. I pressed her firmly to my heart. Then I calmly looked the leader straight in the eye. Without hate or fear, not asking for pity, I whispered, "I am ready. Do what is your duty."

The leader watched me and Martina. His expression changed from one of strict determination to one of love and kindness. Was this not the same warm look I had seen in Ulrich when he left me for the war? The thought of Ulrich overwhelmed me. Wasn't that tall leader with the gentle blue eyes Ulrich saying good-bye to wife and child?... The leader spoke some words to the soldiers, and they lowered their rifles and turned aside. He came to us with the interpreter and said, "The child saved your life."

I broke down. My knees collapsed, I was trembling from head to foot, and my teeth were chattering. I couldn't stop it, I was unable to speak. I felt completely empty. The leader tried to calm me down and offered me a cigarette, but I wasn't able to take it – my hands were shaking too much. So he took his burning cigarette out of his mouth and put it in my hand. Without thinking, I crushed it in my hand. When I felt the sudden searing pain in my palm, I became immediately calm and stopped shaking. The soldiers began to loot our belongings, but the leader stopped their plundering and commanded them to leave. We were told to return to Cracow the next morning. Greatly relieved, we saw them rowing away.

Back in town we heard about the terrible things the soldiers had done to our neighbors. The brutality and cruelty suffered by the young women and girls is not to be described in words. As conquerors, they had been encouraged to do what they liked with us women to take revenge. And they were merciless. It is true that the Russians suffered terribly under the Germans, though, and for that they hated us very much.

Once, I was forcefully pulled down to the ground by a drunken soldier. I shoved him away with all my might. His face, so near to mine, was coarse and full of lust, and I was overwhelmed with horror and disgust. I wanted so very much to hate him, to curse him, but suddenly I saw in him the suffering of all mankind. A deep feeling of great pity overpowered me, a feeling I had never known before. I couldn't hate or curse him. Was he not also a victim of this dreadful and merciless war? My resistance vanished. I was calm, and a peace, not of this world, came into my heart. The soldier was astonished. His arms, which were holding me as tight as iron wires, loosened a bit. In that second I was able to free myself and I dashed away. Furiously he shouted, "Frau komm! Frau komm!" (Every Russian soldier knew that much German.) He shot after me, and although the bullets hissed around me, I was able to escape.

I ran into a nearby forest, where I met five women. We all hid, and when we came out from behind the trees I was sure they would welcome me, glad that I had escaped. But what a shock to see how angry and furious they were! With faces full of hatred, their fists threatening, they surrounded me. "Why did you run away? Why did you not stay with him? Those bullets flew at us, too. How easily we could all have been shot!"

I wanted to shout back, but again I felt something different flood over my heart. I was sorry, really sorry, for these women. I could understand their anger. They only wanted to live, wanted nothing else but to be alive.

On another occasion a soldier forced his way into my room at night. As I cried, "No, no, no!" I suddenly heard a man's voice behind me. He was another refugee, who lived next door with his young wife. He took the soldier by the arm and finally managed to persuade him to leave. The next morning when I thanked him, he said, "You helped my wife once, too. Now it was my turn to help."

Then I remembered how his wife had come running upstairs, trembling with fear. She had seen the soldiers coming into our house, and her husband wasn't there. With all my might I had pushed her under a very low sofa standing against the wall, and I squeezed under it, too. A Russian entered the room, and in spite of the fact that the sofa was shaking visibly, he left again almost immediately. This poor woman was so afraid that she had completely lost control over herself. She was very embarrassed about that, but I calmed her down and said, "Weren't we lucky!" And we laughed together.

Later, her husband told me where he had taken my night-time intruder: to a woman who sold herself to earn food for her children. I knew that woman – a mother of six girls. She looked out of the window and waved to the soldiers as they passed by. Everyone gossiped about her because of that. But when someone told me what a bad mother she was, I said straight away: "Why do you blame her? Give her food for her children." She was a refugee and she possessed nothing, only six children – and those children needed to eat every day.

That wasn't the last of my frightening experiences. One day a young Russian soldier was furious with me because I had left the queue, and he put his pistol up against my right temple and pushed me forward with it. Silently, without any resistance, I let myself be pushed. I felt the cold, round rim of the pistol on my temple. Suddenly he shot twice in the air, and then ran off, laughing at me devilishly. Even a long time afterward, whenever I thought of that event, I felt the cold pressure on my temple...

Some months later Ulrich came home. He had been released from captivity early because he was an agriculturist, and Germany was in great need of skilled men. On his way, searching for us, he had seen and heard all that had happened to the women. He became more and more fearful as he approached Cracow. How thankful and glad he was to find that his wife and child had survived, and that both of them were healthy in body and soul. After two days of rest in Cracow, we journeyed many days on foot back to Eisenach. All the bridges over the rivers were destroyed, and we could cross only by ferryboat. For that we needed permission from the Russians.

They said they would give us permission only if a young woman would come first into their hut. The Russians were drunk and slept all day long in that warm hut. We sat day and night on the bank in the cold rain. After twenty-four hours the young daughter of an old, ill mother went into the hut. She thought her mother would die if she had to wait one night longer in the cold and rain. She went out of love for her ill mother. Who can judge if she did the right thing?

JANE LEARY NEVER SAW war firsthand, but she remains scarred thirty years after the death of her brother John Dennis. As her story shows, there is a long "tail" to the death of even one soldier: a sister, a brother, a wife, parents, children.

In the sixties I was a hippie. I got pregnant and I wasn't married. I decided I didn't want to be with the father, because he was doing drugs. So I gave the baby away for adoption and came home to a father who was dying of lung cancer. I didn't have time to grieve the loss of my child when I lost my father. Then my brother was drafted. He graduated from high school in 1967 and went in April '68. And because my father was a veteran of the Marines, he joined them.

I was really upset with him, because if you were in the Marines you were guaranteed to go to Vietnam. If you were in another part of the service you might get lucky. But he joined, and my father's death had a lot to do with it, because he felt he had to be like his father. If Dad had been alive he might have tried – I'm sure he would have tried to talk him out of it.

My brother went to Parris Island and then Lejeune, and then home for thirty days, and then Pendleton. I drove him to the airport. It was me and my younger brother – my mother didn't want to go. The car was quiet on the way over. Nobody talked much.

I was a secretary then, and I used to write to him every day. He went to boot camp in April, and he left for Vietnam on September 5, 1968. He called home right before he left. My mother wasn't there. I could just hear the despair in his voice. He didn't complain, but it turns out – which I didn't know at the time – that he had called some of his friends the same day and said that he didn't think he'd be back.

One weekend in April I drove up to the home of a friend in New York. His cousin was getting married and he asked me to go to the wedding with him. It was on a Sunday, April 27, and little did I know that it was the day my brother was killed. I stayed overnight in New York and drove home on Monday. I was in a strange mood. I'd had a nice weekend, but I just couldn't figure out what was wrong. I went to work the next day, and I couldn't concentrate. I must have been acting differently, because I remember having lunch with two of the girls I worked with, and they kept saying, "What's the matter with you? You're awfully quiet. Didn't you have a good time?" And I said, "Yeah, everything's okay." And then I can remember trying to think I would write him a letter, and then I thought, "No, I better get my work done first, since I missed Monday."

Well, we came back from lunch and I'm sitting at the typewriter. It was just inside the door, in the reception area. My boss came in and threw his hat on the chair and he went back towards the lab instead of going to his office. I remember thinking, "That's weird." Then the next thing you know, I'm looking up from my typewriter and one of my father's best friends and one of my mother's best friends are standing there. So I know that something has happened – but I really didn't think he was dead.

I remember I was so upset. Adolf was holding me, and he couldn't even talk. I kept saying, "What happened? What happened?" And Mrs. Powers was facing me, and she just said, "He's dead." Then the doctor came in and shoved a needle in my arm…

It was two weeks before I returned to work. There were all these people at the house. My brother had been an outstanding athlete; he had gotten the Brooks Irvine award at the high school, which is the highest honor you can get, and he was voted best-looking in his class. He had been a punter for the Colonial Conference Champion Football team, undefeated. He had gone all the way to the state wrestling championship. He just was really very well liked. The street we lived on went right through town, and on every block there were three or four flags flying at half mast.

Jim came down from New York for the funeral. He'd been shot down in Vietnam, and he'd lost his arm and had badly injured both legs. But he was able to walk in casts. He sang at my brother's funeral, and I don't think there was a dry eye in the church. Sister Wilhelmina, who was the second-grade teacher from the parish church, had all the second-graders out lining the street along the curb. When the hearse went by they saluted. We were lucky enough to get one of the Marine pall bearers, and so they did a gun salute at the cemetery. I still have one of the shells. It seems like I'll never forget him... O God, I was close to Denny!

HASSAN, A VETERAN OF the Iraqi Army whom I interviewed for this book in December 1997, joined in the uprisings in 1991 and spent the next five years in Saudi concentration camps before finally emigrating to the United States with the help of UN officials.

When Iraq entered Kuwait, I was a soldier... Almost all of the young people in my country were in the Army at that time, because we had been at war with Iran for eight years. Most of us never agreed with the invasion of Kuwait, but we were scared to protest. People were forced to fight or die; they didn't have a choice...

I decided not to participate in this war because I felt that, either way, I would die. So in October 1990, I escaped from the Army and I went to my city. Nearly half of our soldiers escaped and hid with their parents... I stayed with my cousins until six months after the war.

I remember the first day of the bombing. It was January 16, 1991. The Americans attacked Iraq in the night. All you could see was smoke, and all you heard was the noise of the airplanes. First they started with airports, phone centers, media, and communications. My city is a hundred miles south of Baghdad, but the sound of the bombing was loud. Between my city and the capital, there is the biggest arms manufacturer in Iraq. They bombed these places, and it felt like an earthquake.

They continued bombing for fourteen days or more. In two days Iraq lost everything. There was no rescue, no resistance from Iraq against the airplanes, no missiles. After three days, I had to go to the capital because my grandmother forgot her medicine, and I had to go to bring it to her. It was a risk, but I felt I had to do it. I went there, and I found too many missiles around the capital. It is the truth, I saw them by my eyes. They looked like shining big airplanes or something. They fell on oil companies, and they burned for a whole week.

Everything stopped in my country: no electric, no water, nothing. People depended on what they saved. It's a custom in my country that we say thank God for giving us what we need. We save every month, as a custom. Not like here in the U.S. So people lived on what they saved in their houses.

The only news we could get was from Saudi Arabia. They reported that we were hiding weapons in the schools, so the schools became a target, and they didn't care. They bombed the schools, which are surrounded by houses. Too many houses-full of people were killed for this…

Then the uprising happened. People were very angry. They were dying anyway, so why not try to change the situation? I lived in hell with this government…

You might say the uprising happened by accident. There was no agreement about it beforehand. Everyone felt the same way, so we started to attack government centers. Then President Bush called on the Iraqi people to rebel, and he promised to support us. But then, when we made the uprising, the Americans let the government crush us. We were between two powers. The Americans said, "We support you," and nothing happened. All they did was support the government for no reason. They said they were enemies, but they looked very friendly in that time. They allowed our government to use helicopters, against the United Nations law, to crush the uprising. They said, "Now only helicopters can fly inside Iraq." They played games on the people…

Our government arrested thousands of people from the cities, even if they hadn't participated in the uprising. They took all the young people away in buses and took them to big camps outside the city. They executed them. And the Americans didn't want to help us. They said, "We are against Saddam," but the truth is that they are against the people of Iraq.

I think the United States doesn't agree with Saddam Hussein, but they want to keep him in power because if the situation changed in Iraq, they would lose their influence...

Iraq lost everything after the war. Saddam Hussein signed a blank check to give them everything they wanted, so why would they want to replace him? It's a political agenda. Because I would feel bad to treat some animals the way the Iraqi people have been treated. So if the United States is concerned about human rights, they would change what they are doing. Because they know exactly. They know that even if they keep the sanctions going for a long, long time, Saddam will not fall.

Who lost his money, the Iraqi people or Saddam Hussein? Who lost his food? People or Saddam Hussein? Saddam gets goods from Italy and Britain, the best furniture, the best things in the world. His wife is the richest woman in the world. They didn't lose anything. But the people, they lost. They lost before the Gulf War when they fought Iran. They lost after the Gulf War. Somehow I don't think there is anything to do with human rights in this. And the big people in the United States, they know the facts exactly. Do you think there is no light in Saddam's house?

I couldn't stay in Iraq. They took us to the border between Iraq and Saudi Arabia, and they put us in the desert. For six or seven months we were under United Nations jurisdiction. They put us in camps built for prisoners of war, in tents. It looked temporary. There are no cities around, no fresh water, nothing. No utilities. Then after eight months, the United States Army pulled out from Saudi Arabia, and the Saudi Army took us. They thought we were prisoners of war; that we had fought against them. They hated us.

I don't know the expression for that kind of hate. And they wanted to kill us, that is the truth. Too many times they attacked us with weapons inside the camps. Some people lost their kidneys because they got shot. Some people got broken legs, hands. It happened too many times.

When the United States Army was still there, they took people from the uprising, who had been shot or injured with chemicals, into their hospitals. They treat us good, I don't deny this because it's the truth. I felt that they gave us the right to talk, to be safe. But then, when they left, people were thrown into the desert for no reason. That's where we lived for five years, and no one talked to us. No health care, no fresh water. People forgot about us...

It was like a fenced farm without trees or animals. They put us there without any utilities. They brought us salty water, which we couldn't drink. Most people had diarrhea for a long, long time from this. When we asked why, they said it's expensive and we are in the desert, so we can't bring these things.

Then, after two more years, the United Nations came to us. I came here, in October 1995. There are still 4700 refugees of the uprising in the camps...

People here in the United States live very well. There are big problems in the rest of the world, but people here don't know really about them. They judge by their government's opinion or by the media's opinion. And they don't care. It's very busy, and most people try to relax after work. I mean, they just take care of their business.

ALEXANDER RIVAS WAS sixteen when I first met him in 1992, and I stayed up till the wee hours of the morning listening as this young El Salvadoran told stories of a country torn by war for twelve long years. His comrade José Peña, a serious young man and a powerful singer, told of the terrible atroci-

ties inflicted upon his fellow peasants, and of his vision for "one country" of peace for all nations and races.

When the revolution began, Alexander was only a small child and lived in an agrarian commune formed by the peasants in response to the growing economic and social crisis. In these "base communities," individual needs could be carried by many and the tide of U.S.-sponsored military oppression faced together. But the communities were soon perceived as a threat to the government and endured a reign of terror. Soldiers continually confiscated food and supplies, harassed and beat the peasants, and shot them when they opposed. Alexander's mother died in childbirth due to inadequate medical care, and he told us of countless others who had died from impossible living conditions, or who had simply "disappeared" after being apprehended by soldiers. One afternoon, just days after another round of interrogations and pillaging, Alex went out into the fields to tend to the community's crops. As he bent down to dig in the soil, he was blinded by a brilliant flash as a violent explosion hurled his body into the air. Both of his arms were blown off above the elbows and one of his eyes was irreparably damaged. The soldiers had planted mines in the peasants' fields before they left in order to "teach them a lesson."

When I met with Alex in 1992, he was wearing an artificial eye that had to be taken out at night. He needed help to eat and bathe, but he was able to hold a phone or rub his eye, which seemed to irritate him continually. I was moved by his attitude to life, which was cheerful and enthusiastic. A bit of a jokester, he was nearly always smiling, nudging his companions with his shoulder or the stump of his arm, and laughing.

One cannot meet a person like Alex Rivas without being angry, despite his cheerful demeanor, at the reckless and callous manufacture of "anti-personnel devices." The United Nations estimates that there are more than a 100 million mines deployed in 63 nations worldwide. Every week, approximately 500 people, nearly all civilians, are killed or maimed by these "devices." They have killed or maimed more people than all nuclear, biological, and chemical weapons combined. It is estimated that it would cost from 20 to 30 billion dollars to disarm all the landmines currently deployed, and even then there are an additional 100 million more stockpiled around the world.[22]

At the time of writing, landmines continue to be manufactured by the United States and several other countries in defiance of both the 1949 Geneva Conventions and an International Campaign to Ban Landmines, which is endorsed and supported by forty-one countries.

The United States has offered plenty of rhetorical support for the ban, but delays and obstacles continue to arise. A recent UN conference was deadlocked partly because the U.S. wanted to explore conversion to "smart" mines which could deactivate themselves, and President Clinton's much-touted call for the elimination of all "dumb" mines by 1999 is a meaningless political ploy: companies stopped making such mines years ago, and existing stocks were already scheduled to be destroyed.[23]

THAT CHILDREN SUFFER terribly in time of war is no surprise, and yet I was astounded to uncover a newsletter put out by Washington's Center for Defense Information that detailed the particularly gruesome plight of child soldiers. As many as a quarter of a million children worldwide have been conscripted to serve in armed conflicts; some of them are as young as five. In Mozambique, recruiters have hardened children by forcing them to kill people from their home villages. In Peru, the rebel group Shining Path has forced young children to eat the body parts of those they have killed. Such abominations remain largely invisible because they occur in remote areas, far removed from media scrutiny. And although one could argue that the phenomenon is not new, there are reasons for its recent escalation.

Involving children as soldiers has been made easier by the proliferation of inexpensive light weapons. As recently as a generation ago, battlefield weapons were still heavy and bulky, generally limiting children's participation to support-roles. But modern guns are so light that children can easily use them, and so simple that they can be stripped and reassembled by a child of 10...

The very high proportion of children in the armed forces of El Salvador during the 1980–1992 civil war suggests this was a routine occurrence. Of the approximately 60,000 personnel in the Salvadoran military, ex-soldiers estimate that about 80 percent, or 48,000, were under 18 years of age.

Quite often child "recruits" are arbitrarily seized from the streets or even from schools and orphanages. Press-gang tactics were prevalent in Ethiopia in the 1980s, when armed militias, police, or army cadres would roam the streets picking up anyone they encountered. Children from the poorer sectors of society are particularly vulnerable to this tactic...In Burma, whole groups of children from 15 to 17 years old have been surrounded in their schools and forcibly conscripted. Children are also recruited from refugee camps and forced to join armed opposition groups in their country of origin or the armed forces of the country providing asylum.

In addition to being forcibly recruited, children also voluntarily present themselves for service. It is misleading, however, to consider this "voluntary." They may be driven by cultural, social, political or, more often, economic pressures. Hunger and poverty often drive parents to offer their children for service. In some cases, armies pay a minor's wage directly to the family. Children themselves may volunteer if they believe that this is the only way to obtain regular meals, clothing, or medical attention. Some parents encourage their daughters to become soldiers if their marriage prospects are poor.

Some are persuaded to join by propaganda and religious fervor. For example, the marching chant of a column of 15,000 Iranian children on their way to the front during the war with Iraq was, "Come on, come on, plunge on. Those who step on mines will go to paradise." It is said those children were sent across minefields ahead of more valuable, trained adult soldiers…

Once recruited as soldiers, children generally receive much the same treatment as adults, including often brutal induction ceremonies…Even those who start out in "support" functions cannot escape exposure to the risks and hardships most often associated with combat roles. Children often serve as porters, carrying heavy loads up to 132 pounds. Children who are too weak to carry their loads may be savagely beaten or even shot. Children are also used extensively as messengers and lookouts…In Latin America, government forces reportedly have deliberately killed even the youngest children in peasant communities on the grounds that they, too, could be "dangerous."[24]

Le Ly Hayslip, who survived Vietnam, said it best: "Children and soldiers have always known war to be terrible…"

One question that always, without a doubt, crops up when people know you have been to war is, "Did you kill anyone?" Even if they don't ask directly, you know that's what they really want to know. It's one of the hardest questions to handle. You are torn between honesty and lies, between wanting to share with the questioner the horrors that accompany the taking of another life, and brushing the whole thing aside because you don't want to talk, because you don't want anyone to know what you've done. But you can't even be selective about who hears lies and who hears truth, because deep down, you really don't want to hear any of it yourself.

Andy

I grew up with a father who was an ex-serviceman. All my childhood, it was "military this, military that." He was full of stories, though he'd never actually been in combat. By the time I was twelve my mind was fully made up: I really had little real intention of doing anything with my life other than joining the forces. So on January 9, 1973, at the ripe old age of sixteen and a half, I signed up with the British Army. My father was as proud as punch, but my mother, being a mother, was totally devastated by the whole thing.

Looking back, it's difficult for me to justify my feelings of anticipation, of wanting to kill. At that time, television was full of the Vietnam War, and this helped psych me up. I think it all seemed like Hollywood. I had no possible way of knowing what happens to the minds of those who are faced with death, killing, destruction, the burning bodies of kids…

I joined the Junior Leaders in North Wales as a training candidate for the Parachute Regiment. By enlisting as a potential for the Paras, it was made all the more apparent to me that I would be expected to go off to war at some time in the future. That's why I joined, I was told: if I'd not wanted that role, I should have gone into the Medical Corp or the Royal Engineers, to learn a trade. Killing was to be my trade. And that's the way I wanted it to be. I wanted to be a soldier, if only for my dad.

During my training the Parachute Regiment's 1st Battalion was in Northern Ireland, which brought the reality of active service closer to us. Also, one of the platoon commanders stationed with us had, it was rumored, led a Para platoon in Northern Ireland during the notorious "Bloody Sunday" massacre. This was sufficient to charge the overactive imaginations of young soldiers and added emphasis to our already very gung-ho training. Airborne forces are Special Forces – the best of the best – and we were constantly reminded of this.

Basic training as a Junior Soldier is no different from what someone who joins the military as an adult goes through. It just takes longer. Having enlisted without any formal education, I was able to sit some examinations towards promotion. But everything about my training centered around killing other human beings. My days rotated around explosives, shooting at people, blowing people up, hand-to-hand combat, taking lives. Very quickly, you have no other thoughts in your head other than to go out and do the job that you've been taught. Because that's what it becomes: a job.

But it wasn't long before I became very bored with Junior training. In an attempt to be transferred to the Parachute Regiment depot, I went AWOL. The stunt worked. I finished my basic training at the depot in Aldershot and was posted straight out to Northern Ireland with my new Battalion, 2 Para – a present for my eighteenth birthday.

This was my first active service tour, and I was decidedly the new boy. I was surrounded by others who were on their third, fourth, even fifth Northern Ireland tour. So I was "on trial," placed in situations to test my courage. I soon learned not to show fear, and after a short while I developed a very tough persona.

Northern Ireland fed me everything I considered I needed to become a bona fide serviceman. There was the chance to be fired at, and to return fire, which always brought with it the most incredible feeling of adrenaline rush. The closest I came to being hit was while on gate duty at the company's base off Ardoyne and Shankill Road. As I and a fellow soldier stood at the wall, we heard gunshots, and then bits of concrete were flying everywhere. They literally peppered the wall right beside me.

There was no time for fear – we just returned fire. I couldn't identify who was shooting at me, but I could see who I was shooting at. (The follow-up patrol found empty cases in our target area from the enemy weapons, but they didn't find any bodies. No doubt if they had found a body, they wouldn't have found any weapons. In those days, whenever someone was shot, others would come and take the weapons away, leaving the soldier faced with a murder charge, but even this possibility didn't make much difference to me at the time.) I didn't care who was shooting at me; I was only interested in identifying a target to shoot back at. That was the mentality I developed. I had several instances like that in Northern Ireland, each one greeted with the same amount of "after-enjoyment." Really, there's no other word for it.

On one occasion, a friend of mine was dragged in from the street, having been blown apart by a car bomb. He survived, but he'd damaged both his legs and, after his recovery, was transferred out of the Parachute Regiment. I felt sorry for the guy, but I think at the time I felt more sorry for him because he had to transfer from the Battalion than because he'd been injured. Getting injured in service was acceptable to me. We were told before we went, "You could possibly lose a friend. You may lose your own life" – whatever. This was just to keep you constantly on the ball, to ensure that you were doing what you were employed to do, which was to patrol the streets and protect the "innocent," though we never really believed there were any innocents.

On patrol, I'd be constantly searching for sniper sites, looking for missing tiles on the roof, missing bricks in the bricked windows of derelicts, checking for holes in walls, open windows to darkened rooms. I was determined that if there was going to be any action, then I would be the one to identify the gunman. I would be the one to shoot the gunman, and I would be the one who came home from Northern Ireland very happy. There was nothing personal in it whatsoever. It was a job. This is what I'd been trained for, this is what I'd joined the Army for. As far as I was concerned, I was in Northern Ireland to kill the IRA, and that was it.

Really, though, aside from a few flash incidents like the ones I've described, life in Northern Ireland was generally quite boring. Even though I knew the dangers, I never actually felt any real feelings of fear while walking the streets. If anything, I felt stupid. I think my training had an awful lot to do with my ability to cope, both because of its intensity and the way it instilled hatred of the IRA.

That was my first experience of Northern Ireland. I returned in 1979 for a second tour and was attached to an intelligence section in Lisburn, which meant plain-clothes duties. And that tour was the first taste I had of the reality of the situation I was in. Walking around the streets of Belfast in civilian clothes was particularly frightening, partly because I knew I looked different from everybody else around me. Long hair was in fashion then, but I looked the typical soldier: short back and sides. As far as I was concerned, my hair couldn't grow fast enough.

One evening I was provided with an opportunity to view life from the other side, as it were. I was walking alone in the Shankill Road area, a place I admittedly had no business being (the military authorities were quick to point this out later). I was carrying a concealed weapon. A friend of mine was out with his patrol on this particular evening, and he stopped me. That was frightening: I wasn't worried about my colleagues, but I feared that if they were overly social with me and somebody took notice, I might not make it home that night.

I told them to search me. "Yes," I said, "I've got a weapon. Just treat me the way you would one of the Micks." Well, they took me at my word – they kicked the hell out of me!

That experience made me look at how I had treated the Irish during my first tour. I had been no different. If I couldn't shoot them, I would kick the hell out of them. I think this was where I first started having little niggly doubts about military service.

After that, I no longer liked the reasons for my being in Northern Ireland, nor did I like what I was expected to do while I was there.

When that tour was up I returned to Aldershot, and not long after that I married my first wife. We'd been together for less than nine months before I was posted to Berlin on a two-year stretch. She moved to Germany with me. Those years in Berlin allowed me time to reflect on why I was in the military, and eventually I decided it was time to get out. I think the experience of being on the receiving end of British Army justice in Belfast remained with me right through my time in Germany, and probably influenced my decision to get out.

I finally bought myself out of the army in 1980 (I had originally enlisted for twenty-two years, with options to leave at various intervals; because I left between intervals, I ended up paying something like £400, which in 1980 was a fair bit of money). My wife and I were going through marriage problems, which sort of gave me a legitimate and "honorable" way to leave without stating my reasons. Military life is never really compatible with married life and many a soldier has left with this excuse.

So I followed my wife back to England as a civilian, back to the area in North Wales where my initial military training had taken place, and I got into an on-the-job training program at the local hospital, working as a medical technician in an operating theater.

Shortly after I left the army, my battalion returned to Northern Ireland for a two-year posting. During their tour the Warrenpoint bombing happened. The IRA planted a mine that wrecked a couple of vehicles and took the lives of fourteen of my friends. Some were guys I'd been very close to. We'd spent nights out partying together, they'd babysat my two kids, enjoyed meals at my house…I can now say that Warrenpoint was the turning point for my life and my marriage. Slowly the guilt trips set in: I had left the regiment before my time, and as a result, I wasn't with my mates as they died. I tried telling myself that my presence wouldn't have changed anything – I would probably have wound up dead, too – but just the fact that I was not there meant I couldn't forgive myself.

This caused major problems in my home, and is probably what contributed most to the complete collapse of my marriage. It was very unfair on my wife, but it was a case of "I only got out because of you, and now I'm out and my friends have all been killed." I really took it to heart – just total guilt, and it wouldn't go away. It cost me my family: I lost my son and my daughter as well as my wife. My guilt over losing them lingers on, but how could they or anyone possibly understand?

The day finally came – I think it was in late '81 or early '82 – when I simply walked out, got into the car, drove to Dover, got onboard a ferry, and headed for Europe. From that time onwards, I went on sort of international walkabouts, constantly touring abroad and burying my sorrows in anything that could provide an escape. And that was it. I didn't come back to England for any substantial period for the next twelve years or so.

The longest I stayed in any one place was in Tenerife, that sunny little island off the northwest coast of Africa. I wasn't really doing anything with my life there, just soaking up the sun, working odd jobs in the local hospital. One night I was drinking and chatting with a bloke in a bar who told me he was involved with a reactionary force that was patrolling for the Red Cross people around the villages in Mozambique. He was offering me a job, and it all sounded pretty good to me. He'd give me a gun, he'd give me a Land Rover, and all I'd have to do was follow the Red Cross around and protect them if we ran into any trouble. So the next thing I knew, I was bouncing around in a four-seater plane, headed for Africa...

As it turned out, what I'd been told was far from the truth. The men I was to work with were actually a mercenary group employed to fight against the Renamo guerrillas, a group who, with the backing of the South African government, controlled much of the land in Mozambique. They were fighters against the government in the long civil war that had been raging for several years before my arrival and continued after I left. So in effect, I was to fight against the South African government.

There I was: no entry visa, no indication of how or when I had got into Mozambique; the last stamp in my passport was from some months before, in Tenerife. If I'd been arrested, my role would have been obvious. And I knew what I could expect after that – these countries are not in any way particularly hospitable to those such as I.

To this day I don't know who my employer was. I was attached to a group of men and was told I'd get £400 a week, and that was it. We had enough money to keep us happy with cigarettes, beer, and other little social amenities. But I'm sure I didn't get my full whack as far as money was concerned when I finally left.

I hadn't even been there a week before we went out on my first patrol. To this day I still believe it was designed for no other purpose than to try out the "new boy." I don't believe that the men we took out had done anything, certainly nothing against us, but orders had come through saying we were to go down to a certain clearing and eliminate those men. Which is what we did. At the time I was quite – and I use this word very reservedly – happy with the fact that I could go in and shoot somebody. For me that was sufficient, that's what I was being paid to do. I could handle the killing. But over the course of the fifteen or so different patrols I was involved with, I saw a lot of blood, a lot of death, many horrible things. And I really couldn't cope with what the others were doing.

There were only nine of us with a few blacks, but we would actually go into small homesteads and completely eradicate them with gunfire and fire. We would be told to go in and clear an area, because there were these guerrillas in hiding there and being looked after by the villagers. We would just go in and clean out the whole village. There were very few men there, because they were all away fighting. The women stayed at home, tending goats, crops and homes while caring as best they could for their children. Half the people would run away into the trees, or whatever, and the other half would die.

There were occasions when we went in against specific rebel groups. This didn't bother me so much, because they would have quite easily come into our encampment at night and killed us where we lay. But the women and the children and the villages we destroyed, it was just horrendous.

Mercenaries are animals. They've got no respect for providing somebody with a quick death. They are torturous creatures and take great pleasure in inflicting massive amounts of pain on individuals.

I've seen men saw people's hands off with machetes, for sheer enjoyment. I couldn't comprehend that, but I certainly wasn't man enough to speak out against these men or to question why they did these things.

I've seen people strung up by their feet, their skin scored so that the blood drips down. Skin is like an orange peel; you can peel someone's skin off very easily, very quickly. And then the ants and jungle insects would finish them off. A long, slow, very painful, horrible death. And the smell...

When I was on my own and had the chance, I would try to make sure death was quick. If I was with the others, I felt compelled to go along with what they were doing. And that's a very horrible situation to be in, because you're actually fearful for your own life at that stage.

There's a very fine line between being able to kill and being a killer. I don't actually regard myself as a killer.

One time we entered a village just after it had been attacked by the guerrillas, and I was immediately set upon by a huge, fat black woman who was screaming and yelling at me, and beating me in the face. I had no idea what it was all about. Anyway, I got one of the guides to explain what the story was. It turned out there was a British mercenary who'd been with the guerrillas whose nickname was Elbow. He'd got his nickname from his activities. He would take young children – he seemed to prefer babies – and, while restraining them down on the floor with one hand, he'd crush their heads against the pavement with the elbow of his other arm.

I found this unbelievable and started asking questions. Eventually, when I concluded I had no reason to doubt the facts, I decided I couldn't leave Mozambique until I had done something about Elbow.

What this man's background was, I didn't know. All I knew was that he was a British mercenary and he was renowned in the area for his "hobby." But word got round very quickly that I was asking questions. About two weeks later, I was woken up in the middle of the night; somebody had come to find me. They'd found Elbow's location with his section.

The people I was working with were already aware that I was looking for this bloke and what my business would be when I found him. They seemed quite happy to let me go and gave me some money that they owed to me – something like £1500, which wasn't an awful lot for three months work. So that same night I went with the men who had come to find me to their village, taking a vehicle loaded with provisions.

We spent three days preparing for the patrol, then moved in just before dawn. Each man in Elbow's section died that morning – every man, that is, except for Elbow. I stood over him while he slept and fired the first rounds of the attack. For some strange reason I wasn't happy to kill him. I don't know what overcame me, but at that moment I just wanted to hurt him really badly and leave him with something to remember, or regret for the rest of his life. So I totally destroyed his arm, firing a burst into his elbow.

I killed two people that night, and we all made it out of the encampment and back to our vehicle. We couldn't have gone more than three miles when we hit a mine, and I lost the three men who had fought with me and helped to destroy Elbow's patrol. One bloke had his stomach ripped out by the blast, another lost both his legs, the other one – I don't even know what killed him; he was just dead.

As I pulled myself up from the road a sharp pain hit me in the side. Reaching down, I found a piece of metal embedded in my body. I knew that to remove it would risk allowing air into my chest cavity and cause one lung to collapse. So I stood holding it, looking for something to plug the hole. I found some mud, removed the metal, and patched myself up as best I could. I took my weapon, maps, water, and some ammunition and spent the next four days travelling back to my friends' village.

There I had to face the women, telling them that not only had I not killed the British mercenary who had murdered their children, but I was also responsible for the deaths of three of their men-folk. I don't honestly know which was harder to say. And I still don't know why I didn't kill Elbow.

I remained there in that village recovering from my injury for some weeks before making my way back to Tenerife and the Canary Isles, where I just wasted a further three years or so. I literally dropped out of society. I hit the drugs, hit the drink. I lived the life of a beach bum, selling time-share, running bars, and any number of things that would keep me wasted and out of touch. I was happy; memories of my four months in Africa were completely blanked out. I didn't want to know about it, didn't want to hear about it. I was simply unwilling to believe that I'd been involved in so many horrible things, couldn't accept that I had been involved in some of the things I'd done.

During my time in Tenerife I met the woman who later became my second wife, a British girl. She headed home, and we married after my return to England. It wasn't long before she called me at work one day to tell me she was pregnant. I tried to settle into a normal life. I gave up the drugs and went back to work at the hospital, but I continued drinking, even after our son Daniel was born. Before his arrival, I quit my hospital job and got work locally, in Manchester, running an Indian restaurant for a doctor who was out of the country.

And that's what I was doing when parliament first announced that reservists were being called up for the Gulf War. As soon as I heard that, I completely bypassed my decisions never to go back to war following my experiences in Africa. My memory jumped straight back to Warrenpoint, Northern Ireland, where fourteen of my friends had been blown to bits. Here – right now – was my opportunity for redemption, as I saw it: I would go to war with my friends, and if anything should happen, I would be right there. As well as my combat experience, I now had medical skills that I hadn't possessed before. I was as well equipped to save lives as to take them... Before my wife had even come home, I'd phoned the local recruitment office and said, "Here's my name and address. Send me a letter."

I had volunteered. It was not until afterwards that I thought, "What am I going to tell my wife?" I lied and told her that I was being called up, which she accepted for some time – unfortunately, that time ran out. But I was happy. I was going off to war again with British soldiers, with my friends.

Though I still wore my Parachute Regiment beret and uniform, I was recruited as an operating theatre technician and attached to 33 Field Hospital, near the town of Al Jubail, Saudi Arabia. This effectively took away all my reasons for wanting to go to war in the first place, and I tried everything I could think of to be reallocated to an FST (Forward Surgical Team). I must have been off my head, but I really was not happy in a rear position! I couldn't explain my reasons for having volunteered to be in the Gulf, because no one would have accepted my reasoning or rationale.

At the field hospital, I was put in charge of security. I never actually saw the war as such. Neither did I see too many casualties of the war, with the exception of the Iraqi POWs that I watched over and escorted occasionally.

Only the frequent SCUD missile alerts served to remind us that we were at war. They brought an unfamiliar fear to me: the potential of chemical and biological attack. There were an awful lot of unknowns surrounding each alert.

The most I could do at such times was to go around and make sure people were putting their protective NBC (Nuclear, Biological, and Chemical) suits and gas masks on, and to present as brave a face as possible for those less accustomed to such stressful situations.

On January 20, 1991 an alleged chemical SCUD had been launched against us, supposedly contaminating the immediate area around the hospital. I wasn't at the hospital when the news first came in, but as soon as I arrived, I saw the guard were all in their gas masks and full NBC kit. Everyone else was tucked away on alert. I went straight into the duty hut, where a discussion about what ought to be done next was in process. I asked whether they'd done certain standard tests, which they had. "Well," I said, "what about the sniff test?" They hadn't.

Well, I was in the wrong place at the wrong time, as I seemed to be the only one up to date on the protocol. It's simple, really: just expose a human being for up to half a minute to an environment that has already tested positive to some contamination tests for nerve or chemical agents. Within the suspected contaminated area one man is required to remove his respirator while standing two feet away from another man, who counts slowly to thirty, while watching the other for any signs. As the NCO at the time, I didn't feel happy volunteering someone else for the job, so I took off my own mask. And that, for me, was the most frightening and horrible experience of the entire Gulf War, one that still plays on my mind. In that situation, you're exposing yourself to contamination and possible death, while everybody else stays in their respirators and watches what happens to you... Lucky for me, the area proved to be safe.

In the end, this turned out to be one of the few things I was able to do in the Gulf that provided a sense of satisfaction later. During a post-war deep-and-meaningful drinking session, I met a bloke from Bradford who told me that if it hadn't been for me, he would never have kept his sanity through his three months in Al Jubail. One of the female nurses at the base had also said earlier, "You really helped me sleep at night, knowing there was a real soldier in the camp with us." And I had to accept those words of thanks as my redemption experience.

After the war I and some friends drove up through Saudi Arabia into Kuwait City. En route you go through the Basra pass, which is a bottleneck where the road narrows tight. In the final hours of the war the retreating Iraqi troops were caught up in that bottleneck and completely decimated, bombed to bits. Something in the region of nine to fifteen hundred Iraqi vehicles, some loaded up with contraband stolen from Kuwait, had been trapped there. The resulting carnage was akin only to a scrap yard containing not only hundreds of bombed-out vehicles, but also thousands of dead, burned, and putrefying Iraqi soldiers.

We drove into that area, picking our way around bombsites, each of us growing silent as we viewed the scene of utter destruction. Some vehicles were still on the road, but most had been pushed off by the Pioneers. There were bodies of people, still holding the steering wheel, completely charred. The heat had been so intense, their skin just burned onto their bones. And in the silence as we drove through, Elton John's song "Sacrifice" played on the van's stereo.

Try to imagine the smell of tons and tons of rotting flesh, of cordite and scorched metal. Take a piece of pork, set fire to it, and throw it out in the sun for two days. Then pick it up and smell it – and then multiply that smell by two million times.

And the words "it's no sacrifice at all" spilled from the stereo speakers.

That scene will never leave me. I dream about it. It wakes me up. I hear that song and the smells return. I see. I hear.

The war was over. I went home and tried to resume my life. But I didn't sever my military ties completely. I helped organize a reunion of Gulf veterans, celebrating the first anniversary of the war, in March 1992. There in the newspaper was my face, my name, and my phone number for veterans to contact. It was not only Gulf Veterans who picked up on this, however.

At about half past eleven one night, my wife phoned me at work. She was crying. "What's the matter?" I asked. "Who the hell is Elbow?" she sobbed. At that moment, the bottom dropped out of my world. "Don't stay in the house. Wake the baby and go across the road to John's. I'll call the police. Don't ask questions – just go. I'll be there soon."

Elbow had caught up with me. He had phoned my wife – she had no idea who he was; I'd never mentioned him – and told her that he was going to kill her and my son, and then he was going to kill me. His history was sufficient for me to take these threats most seriously.

To cut a long story short, my wife was taken to a safe-house, and I ended up trawling around seedy London pubs, meeting old contacts, and ended up putting out a contract on Elbow for £6000. All I got for my money was a photograph of the bloke in a wheelchair in Rio de Janeiro airport. That was the last I heard of him.

That incident gave me an opportunity to tell my wife things about my past that we'd never discussed before. For a while, things went smoothly, but bit by bit our relationship was crumbling. We seemed to be at each other's throats all the time, quarreling over the littlest things.

Later, I learned that I, like so many other Gulf War veterans, suffer from illnesses linked to a number of plausible causes, including chemical and biological agents, untested vaccines, organophosphate poisoning, trauma, and all the other things loosely grouped under the heading of Gulf War Syndrome. But in those first months after I came home, there was no way of linking any of my feelings – either physical or emotional – directly to the war. Not that I blame the Gulf War for the collapse of my marriage, but I think my experiences and condition were contributing factors.

In the end, my wife moved out, taking Daniel with her, and she got a court injunction to keep me away from him. After that, there didn't really seem to be any reason for me to go on. I got into the car and spent fourteen hours driving around Manchester, seeing nothing, hearing nothing. It was nighttime when I stopped to fill up the tank, and then I headed for the motorway. I was angry, not at my wife or anyone else, but at myself. That anger built as I got on the M1 heading south, towards London, doing about ninety miles per hour. I went straight into the outside lane and picked my speed up to about 120, ignoring the horns blaring around me. I'm not really sure what happened next: one moment I was driving the southbound M1 and the next I was over on the northbound side – still speeding south, laughing at the top of my lungs. I was looking for someone to hit. My laughter soon turned to tears, partly because I could find no one to run into, and partly because I knew that the next time my son saw me, I'd be in a wooden box: a father he would never know, except as a madman who had killed himself and some other poor soul trying to get home.

I reversed off the motorway at the next junction and sat in the car crying for what was left of the night. Why? Why? I kept asking myself. No answers came. Only more questions.

That was my first attempt at suicide, and it wouldn't be my last. For a while, I lived on the streets; I didn't care what happened to me anymore. But over time, and through the help of friends, I pulled myself together and started trying to rebuild my life.

In mid-1995, I began to write my story. I was trying to come to terms with the turmoil in my head and wanted to fill in the gaps in my life – gaps lost through drug and alcohol binges, homelessness and drifting, gaps that lasted for years rather than days. In the process of writing, I came to the conclusion that my life up till that point had actually been meaningless. My only achievements had been to ruin an awful lot of people's lives over the years.

That's what I'm most sad about. I'm not sad that I've taken lives, because I know that if I hadn't, then I wouldn't be alive today. But when you kill somebody, you haven't just taken another life. You've destroyed the life of that man's family, his friends, his children. And you've also ruined your own life, because you can never be the same afterwards. And I don't think I'll ever be rid of those feelings. They're too strong, too deep. They have affected my whole rationale towards life.

I have had my fill of war, of fighting, of all violence. My experiences have left me scarred, jaded. I'm scared of myself. People don't realize how fragile a human body is. I now know what I am capable of. And it frightens me terribly, because I know what I'm capable of doing to another person. I study karate now, because I need to learn control. I am uncontrollable. It's so deep inside me now that there's a point where I can't walk away. I don't go out at night. I can't enjoy myself in the clubs, the pubs; I'm not even able to relax in heavily populated rooms. I stay away from any situations where I might ever have to retaliate. Because I don't honestly believe I would know how far I'd go to protect myself.

Somebody asked me once, "How can you have been through what you've been through and then be scared to go out?" I can't explain it to people. My level of tolerance is so low and my temper and reaction is so bad that I can get angry over the slightest thing. It really doesn't take an awful lot of pushing for that anger to stop being vocal and turn into something physical. And I don't ever, ever want to be in that position again. So I avoid situations that might deteriorate, which basically means I have no social life whatsoever. The war ended all that.

Somehow I've got to make a new life for myself. I've got to put the past behind me – and I've done that as much as I can, but I don't think I'll ever lose the pain over what I've done to other people.

There's the old question, "If you could live your life over again, would you do anything differently?" And some people say, "Hell, no. I've had a wonderful life. I've done a shitload of stuff for other people." Well, if somebody asked me, I'd tell them, "I wouldn't join the Army." And if someone were to ask me for advice about joining the military now, I would tell them, "Go to the movies. Get a video game. Join a sports club. Find something else to use your energy on." Because if you do join the army, at the end of the day you've got to come up with the goods, and unfortunately the price you pay for coming up with those goods is more than I think any man should have to afford.

The personal cost of military training and war is not restricted to the soldiers involved. It touches their families and friends as well. To my score, I've made serious effects on the lives of two wives and four children. My parents never see me; they can't understand me.

At the moment, I live from day to day and from week to week, without any long-term plans. I really don't know which direction I'm going in at the moment. I'm constantly searching, searching – searching for something to do with my life, taking into account that I'm forty-one years old now.

I don't know how much of a grip on life I really have. I know that I have an illness which is similar to other Gulf veterans. I know that this illness is allegedly the result of one or more extrinsic factors during the Gulf War that are not directly related to the actual fighting. I've seen friends that have the same illness, but theirs is much more advanced. I don't know whether I've got a year before the symptoms catch up with me and take over. I don't know whether I've got twenty-five years, or whether I'm going to be dead in ten years, five years, one.

More than ever before, my time matters to me. It matters to me, because I have to make amends for everything that I have ruined. Somehow I've got to get atonement for that, I've got to do some good in my life. But it's very confusing. I've always tried to be there if somebody's got a problem: "Andy, I'm short of cash." "Fine, there you go, there's some cash." "Andy, I've got no cigarettes." "There you go, have some cigarettes." I'm constantly doing things, but you know, there's a limit to how much you can do for people.

They say that time is a great healer for grief. I can only hope that the time for healing is shorter for those who have lost a family member through war than for those who have taken that life.

Let us put up a monument to the lie.

Joseph Brodsky

In the course of meeting with veterans and learning about the wars of this century, I have often marveled at the selective silence that blankets so much of what has transpired. Selective, because there are certain things we are allowed to see and to know. We know all about the impudence of Pearl Harbor and the valor of Iwo Jima. We have seen countless images of the concentration camps: the bulldozers unearthing mass graves, the chimneys of Auschwitz. The Wall, with its 58,000 names, has become a household word. And in recent times we have been treated generously to images of burning Kuwaiti oils wells and acts of Arab terrorism on the six o'clock news. Why is it, then, that we haven't heard the other side?

Why have we rarely seen photos of the Japanese internment camps in the U.S. during World War II, the cities of Dresden or Tokyo after they were carpet-bombed, or the city of Hiroshima in the first hours and days following the detonation of "Little Boy?" A monument in Okinawa displays the names of American and European soldiers killed along with the names of Japanese and Korean war dead. Where is America's "Wall" for the Vietnamese? Why have we never heard of the "smart-bombing" of the Al-Amariyah Shelter, which incinerated hundreds of Iraqi women and children?

Perhaps it's just human nature, plain and simple. We prefer to broadcast our accomplishments rather than our failures and trumpet the injustices done to us while downplaying our own wrongdoing. We hope that covering up the dark parts of the past will help us feel better about our future, and we shy away from the truth, because it hurts.

FORMER MARINE SMEDLEY BUTLER chose loyalty to his conscience over the adoration of a mindless populace. Butler joined the Marine Corps during the Spanish-American War of 1898, earned a medal during the Boxer Rebellion in China, won the Medal of Honor following the capture of Fort Riviere in Haiti in 1915, and was promoted to major general in France during World War I.

As a recent letter to the editor of the New York *Daily News* shows, Butler is still venerated as a shining example of the "few and the proud":

> Our reputation as "the first to fight" and as the world's finest fighting force was built on extreme personalities such as Smedley Butler...Ask any enemy who has had the misfortune to face Marines in battle, if they lived to talk about it, and they will agree...We are *Semper Fidelis* and even our Marine's hymn claims that "the streets of heaven are guarded by United States Marines"...

The effusive manner in which the writer of this letter praises the Marine Corps is hardly shared by Butler himself, however. Despite two congressional Medals of Honor and decades of distinguished "service," Butler later spoke out plainly against the realities of American armed intervention into foreign affairs:

> I spent 33 years and 4 months in active military service as a member of our country's most agile military force – the Marine Corps. I served in all commissioned ranks, from second lieutenant to major general. And during that period I spent most of my time being a high-class muscle man for Big Business, for Wall Street, and for the bankers. In short, I was a racketeer, a gangster for capitalism.

I suspected I was just part of a racket at the time. Now I am sure of it. Like all members of the military profession, I never had a single original thought until I left the service. My mental faculties remained in suspended animation while I obeyed the orders of the higher-ups. This is typical of everyone in the military service.

Thus I helped make Mexico, and especially Tampico, safe for American oil interests in 1914. I helped make Haiti and Cuba a decent place for the National City Bank boys to collect revenues in. I helped in the raping of half a dozen Central American republics for the benefit of Wall Street. The record of racketeering is long. I helped purify Nicaragua for the international banking house of Brown Brothers in 1909–12. I brought light into the Dominican Republic for American sugar interests in 1916. In China in 1927 I helped to see to it that Standard Oil went its way unmolested.

During those years, I had, as the boys in the back room would say, "a swell racket." I was rewarded with honors, medals, and promotion. Looking back on it, I feel that I might have given Al Capone a few hints. The best he could do was operate his racket in three city districts. I operated on three continents.[1]

George F. Kennan, architect and commentator on U.S. foreign policy for almost fifty years, summarized Butler's "American approach" in 1948. As head of the State Department's policy planning staff, he wrote in a now-famous memorandum:

We have about 50 percent of the world's wealth but only 6.3 percent of its population...In this situation, we cannot fail to be the object of envy and resentment. Our real task in the coming period is to devise a pattern of relationships which will permit us to maintain this position of disparity...To do so, we will have to dispense with all sentimentality and day-dreaming...We should cease to talk about vague and...unreal objectives such as human rights, the raising of the living-standards, and democratization. The day is not far off when we are going to have to deal in straight power concepts. The less we are then hampered by idealistic slogans, the better.[2]

Half a century later, Kennan's policy is alive and well. And it is a policy that remains shrouded in a fog of deception and willful ambiguity.

In August 1990, while the U.S. military was winding down its most visible spate of exploits in Central America, and the U.S. government was laying the groundwork for the Gulf War, Senator Daniel Patrick Moynihan observed: "The secrecy system protects intelligence errors, it protects officials from criticism. Even with the best of intentions the lack of public information tends to produce errors; the natural correctives – public debate, academic criticism – are missing."[3]

A 1993 issue of *The Defense Monitor* similarly concurred: "We must...lift the veils of secrecy and get the public involved in decision-making if our democracy is to operate effectively in the post-cold war world."[4]

FEW PEOPLE KNOW the extent of deception permeating the Vietnam War better than Daniel Ellsberg. A brilliant student at Harvard in the late 1940s, he trained as a Marine in the early fifties and then joined the Rand Corporation, a private think tank which advised the Pentagon on strategies and policies of war, particularly the retaliatory strategies of potential nuclear war. His rise at Rand was meteoric, but in the early 1960s he surprised his colleagues by volunteering to go to Vietnam to

observe firsthand the implementation of the U.S. "pacification" program. Remarkably, he ended up in live combat more than once; he has kept a handful of shell casings, which he picked up while still hot and doused in the water of a rice paddy.

A high-level military advisor by the late 1960s, Ellsberg's moment of truth came in 1969, at an antiwar rally in Philadelphia. Thousands of young men were dying for what he knew was a lie, and he had the ability – and the influence – to put an end to it all. His actions would certainly be regarded as treasonous, but if draft resistors were willing to go to prison, so was he.

In his position at Rand, Ellsberg was privy to a top-secret study of America's Cold War involvement in Southeast Asia, commissioned by Secretary of Defense Robert McNamara, which he knew revealed rampant "miscalculation, bureaucratic arrogance, and deception" by U.S. policy makers. The document patterned a history of death and deception, greed and power – in short, it stunningly exposed American imperialism at its worst. With the help of his young children and his friend Anthony Russo, a former Rand employee who'd been fired for exposing torture and other American war crimes in Vietnam, Ellsberg began to duplicate the 7000-page document. He sent copies to antiwar Congressmen and then leaked a copy to the *New York Times*, which printed selections of the so-called "Pentagon Papers" in June 1971.

In the national furor that ensued, Nixon ordered the *Times* to stop publishing the classified papers, and Ellsberg was charged with conspiracy, theft, and violations of the Espionage Act. Nixon secretly directed his henchmen to "incapacitate" Ellsberg, but the plot was revealed by John Dean during the Watergate hearings, and the trial was called off. Ellsberg was free, but Nixon's credibility was seriously damaged. The revelation of Nixon's crimes figured crucially in his impeachment proceedings, which led to his resignation and ensured that the war was never reopened.

I tried to piece together all this history as I waited in my car, in the pouring rain, outside Ellsberg's apartment in Washington. Here was a Pentagon insider who had risked family, friends, and career – in fact, his very life – to stop the war.

Ellsberg gave me a warm reception. He is a man of intense energy, and though he is slightly graying, he gives the image of one still young and vital. His modest apartment is crammed with books: volume after volume on foreign policy, Vietnam, the Kennedys, the Cold War. But these are more than just reading material – in a very real way, they are the chronicle of his life and his work. One senses keenly the presence of history, the fact that here is a man who has seen it all and turned his back on what others embraced. I settled into a comfortable chair next to Ellsberg, eager to hear his story.

What unfolded that afternoon went far beyond my expectations. Ellsberg moved quickly from bouts of passionate reflection to moments of speechless tears, and he apologized more than once for his emotion, noting that his memory is especially acute just now, as he is writing his own memoirs. But the experience was clearly cathartic for him, and I was

later to reflect on the fact that he is still paying dearly for his actions, having lost a lifetime of friends and colleagues in this city that he once tried so desperately to awaken.

In the course of nearly three hours, he told his story – the story of a war, four Presidents, and a very sick nation.

At the Rand Corporation, I worked on what seemed like the most important problem in the world: to deter a nuclear attack by the Soviets. We based our strategies on top-secret estimates of the Soviets' capabilities, and at the time it was believed that they were in a crash program, building up the capability to wipe us out, to wipe out our retaliatory forces, all of our strategic air command bases in one big surprise attack. That was the so-called missile gap – they were believed to be far ahead of us in the number and destructive capability of nuclear weapons, and of course they had put up their first satellite before us.

Everyone I worked with at Rand really believed that this surprise attack would come anytime, not so much in 1958 when I started, but definitely by '59 or '60. It was regarded as a real danger. And I worked night and day, seventy hours a week, to figure out ways to retaliate, but also to deter, to prevent that attack from ever taking place. I should say that my attitude toward nuclear weapons was as abhorrent back then as it is now; that was true ever since Hiroshima. So I was in effect working on nuclear retaliatory capability, and ultimately nuclear retaliatory plans, which might seem a total irony or reversal, but for me it wasn't. The whole point was to prevent any such war from ever taking place. It just seemed that the only possible way to prevent nuclear war, which we thought would come only as a result of a Soviet attack, was to ensure a prohibitive threat of retaliation.

Rand is a private, not-for-profit research corporation that uses foundation money (to start with, Ford Foundation money) to do research in the public interest. In those days we worked almost entirely for the Air Force, and then later for the Secretary of Defense. I was one of the first people at Rand to work directly for the Secretary of Defense, but my work was obviously and essentially for the President, initially Eisenhower and then Kennedy. I was a Democrat, a Cold War Democrat, but my initial work was for Eisenhower.

That's a long story, how I tried to change the Eisenhower plans, which I thought were very dangerous. The important thing is that in September of '61 I learned that the Soviet's "crash program" was an illusion. There simply was no such effort and no such force.

When I visited the Strategic Air Command headquarters in Omaha in August 1961 it was estimated that the Soviets had 1000 missiles. We had only 40 ICBMs and perhaps 120 submarine-launched weapons, and 60 or so intermediate-range missiles in Turkey and England that were within range of Russia. Not to mention 600 B-52s, 1400 B-47s, and a thousand tactical bombers on bases like Okinawa or Korea or on carriers, or in Europe, largely in England. So, 3000 bombers and roughly 200 missile warheads, about 40 of which were ICBMs.

But in September of '61 we learned that the Soviets had not 1000 missiles, which was the SAC estimate, and not even the 120 to 160 that was the national estimate. What they had was 4, compared to our 40 or 200 or 3000, depending on how you figured it. But at a minimum we had a ten-to-one advantage in ICBMs alone, not counting everything else we had. And the 4 the Soviets had were 4 missiles on one pad. They could have hit Washington, let's say, or even Omaha, but that wouldn't have done them any good. They had no capability to speak of at all, and the only force that existed as a first-strike force in the whole world, a force that could disarm any opponent, was clearly ours. So it was a total delusion, and it took me years to realize, on the basis of a good deal of evidence, that our Joint Chiefs had known it all along.

Now the Army and Navy had been questioning those numbers for some years, but those of us working for the Air Force were told that these were treasonous, wishful estimates to keep us from buying the number of missiles we really needed. The Army and the Navy said that the Soviets had only a very few at best, and they were right. The Air Force had lied, and it was a wishful delusion.

I tried to understand how we'd gotten into that situation, how that delusion had arisen, and then I found myself in the middle of the Cuban missile crisis the following year, which really did come very close to blowing the world up, and which again involved a great deal of illusion, on both sides, as to what the risks really were. Obviously, there had been great misunderstanding on both sides, but that's another story...

Anyway, I was offered a job at the Pentagon. I was actually working in the Pentagon already, as a consultant and a researcher, but I was then offered a post on the GS-18 – which is the highest civil service rating. I started at the top, in other words. The only position higher is a presidential appointee, the deputy assistant secretary. So I had the highest civil service super-grade, equivalent to lieutenant general. And I thought that from the inside I'd be able to understand things better and identify these risks so as to reduce them. I was assigned right away to work on Vietnam.

The very day that I started work in the Pentagon was the day on which an attack was supposedly going on, twelve hours away across the globe, in the Tonkin Gulf. An attack was supposedly taking place at that very moment against our destroyers, for the second time, there having been a daylight attack two days earlier. It's not clear to this day whether even that was an attack, although there had been boats in the vicinity of our destroyers two days earlier. Now on August 4, 1964, they seemed to be doing it again: firing on us with their torpedoes.

There were supposedly a lot of torpedoes in the water this time, unlike the first time. But the commodore of the two-destroyer American squadron wanted to wait until daylight before responding; it wasn't clear that there were any torpedoes, and he wanted to check for wreckage. Already by then they realized that most of their reports, in fact all but one – and it later turned out that that one, too, was false – had been mistaken: the sonar man had somehow mistaken the beat of the ship's propeller against the curving wake for torpedoes. So the torpedoes didn't exist.

But President Johnson was unwilling to wait until morning, unwilling, really, to miss the chance of launching an attack against North Vietnam and showing the world, the public, that he was as tough as his opponent, Major General Goldwater. Goldwater was a senator, an Air Force major general in the reserves, who was calling for the bombing of North Vietnam. And, of course, he was Johnson's opponent in the presidential race, and Election Day was only months away. Johnson wanted to show the public that he could be just as tough as Goldwater – just as tough, but at the same time restrained and precise. He would reply just once, with sixty-four planes. He certainly wouldn't undertake an unlimited bombing campaign until after the election.

He knew full well that the circumstances were very uncertain, but he announced that the attack on our ships had been "unequivocal and unprovoked." Even then it was clear that it was not unprovoked, because that very night we were attacking North Vietnam in covert operations. But we went ahead with the attacks, and that was the beginning of the bombing. And then it was all over, until after the election. The President thought the public had seen enough bombing for that period.

Like Wilson and FDR before him, Johnson ran on the notion that he would keep us out of the war. He promised that he would not expand the war, in contrast to Goldwater, who was calling for an immediate widening of the war, the bombing of North Vietnam, and the possible use of nuclear weapons. In fact, Goldwater had called for the delegation of the authority to use nuclear weapons to field commanders, and Johnson turned this into the number-one issue of the campaign, even greater than Vietnam – that this was an irresponsible suggestion, and that as

President he had not delegated and never would delegate his most serious responsibility. He ran as someone to be trusted, someone far more responsible than his opponent. Actually I knew that Johnson had delegated that authority, and General Goldwater also knew, but he didn't announce it because it was a secret.

I was in the Pentagon, and we were making plans to widen the war as soon as the election was over. At the time, I believed this to be a very wrong judgment, also wrong practically. It would not succeed, and it was immoral. Illegal and immoral. I thought it was totally wrong, but it was my job, and so I did it. It's hard for me to explain now why I didn't just quit or expose it. I should have exposed it, definitely, but it didn't occur to me to do it at that time.

So we were now at war. The bombing started early in 1965. And at that point, being a former Marine, I thought, well, now we can't turn back. We can't afford to lose this. At the very least, we've got to fuzz things up if we are defeated. We just can't afford a blatant defeat, for international purposes.

I didn't think the bombing would accomplish anything, but with troops we could definitely avoid losing the whole country. I was not opposed to the idea of putting troops in, now that we were there. So we put troops in.

By then I felt we were getting misled a great deal by Washington, and I didn't want to watch the war from Washington. I wanted to go see things for myself. So I volunteered to go to Vietnam, and I said I would go at any level, any salary that would pay my alimony (by then I was divorced and had two children). But they transferred me at the same level I was at, which was FSR-1, foreign service reserve officer-1. So I got a bonus for going to the combat zone...

I remember asking a radio operator there if he didn't feel like one of the Redcoats. He said, "I've been thinking that all day." It was impossible for anybody who'd been through American grade school to miss the imagery of it. Here we were walking tree lines in somebody else's neighborhood, 10,000 miles from home, big guys weighed down by web gear and helmets and packs and heavy equipment, being fired at by militia, guerrillas – just like the minutemen at Concord. There was the unspoken realization that we didn't have any business being there…

I got hepatitis in the field. I came back, went back to the Rand Corporation, and worked on a study of the war that later came to be known as the Pentagon Papers. The study had been commissioned by Secretary of Defense McNamara, and it examined the history of U.S. decision-making in Vietnam from 1945 to 1968. The reason it ended then, in 1968, was because many people believed the war was really over. The Tet Offensive had shaken public confidence; Johnson had announced his plans to retire, and the bombing had been stopped, at least ostensibly. But the war wasn't over. It had seven years to go. By 1969 it was clear that Nixon intended to carry on the war until he was successful, pretending all the while that he was going to get us out. In other words, the country was being misled in the same way that it had been five years earlier, in 1964. And I started to wonder what responsibility I had, now that I was convinced we were going in a very bad direction.

I should say that by then, by 1968, nearly everyone who had been to Vietnam was disillusioned with the war. They realized that what we were doing could not succeed. Many of them believed that there were other ways in which we might succeed – perhaps by sending more troops or something – but if they had been there long enough, they had also realized that we weren't going to change what we were doing. Therefore there was no chance for success, and that meant that the people who were dying or being killed were doing so without any useful effect of any kind. Nobody was being benefited by it. It was wasting lives, and it was wrongful.

That was especially true of the bombing. It was obvious that the bombing was doing no good at all. My friend Mort Halperin was in the Pentagon then, and in March 1968 he said to me, "You know, there's only three people in Washington who believe in what we're doing." He was speaking about the bombing in particular, before the bombing. We counted: Johnson; Rusk, the secretary of state; and Rostow, Johnson's special assistant. We tried to think, is that really all? We tried to think of anyone else who believed that it was worth bombing, but we couldn't. And Johnson kept bombing North Vietnam until just before the election, even though everybody understood that it wasn't achieving anything.

At that point the bombing had five years to go (Congress stopped the bombing in '73), but we'd already dropped one-and-a-half million tons. And there were six million to go. Four times as many! We dropped two million in all of World War II, including the Pacific. After 1969 there were still three World War IIs ahead, at a point when everybody understood that the war was totally pointless, totally useless – especially the bombing.

Now, two things affected my life at that point. I'd been reading Gandhi since the spring of '68, when I happened to meet people from the Quaker Action group at a conference in Princeton. I had gone there to study counterrevolution, and they were there as nonviolent revolutionaries. So I started reading Martin Luther King's *Stride Toward Freedom,* and Barbara Deming, who wrote an essay called "Revolution and Equilibrium." I read and reread many times a book by Joan Bondurant called *The Conquest of Violence,* on Gandhian thought, which converted me very strongly, very impressively.

Then in late August 1969 I went to a conference of the War Resisters League (they were founded by World War I COs; Einstein was once their honorary president), and in the course of this conference I was induced to go to a vigil for somebody who was going to prison for draft resistance, which was a very unusual thing for me to be doing. There I was, standing in the street outside the Philadelphia post office, passing out leaflets. This was not the sort of thing a GS-18 did. It seemed, you know, rather undignified – giving away your influence and your access in such a ridiculous way, just handing out leaflets like a bum.

Then at the end of this conference I met another young man, Randy Kehler, a Harvard College graduate who had gone on to Stanford but then stopped his studies to work for the War Resisters League. He gave a talk, and at the end he announced that he was on his way to prison for refusal to cooperate with the draft. And this came to me as a total shock. It just hit me that it was a terrible thing for my country that the best he and so many others could do was go to prison. I went to the men's room and just sat on the floor and cried for about an hour and thought, "My country has come to this? We're eating our young. We're relying on them, to end the war and to fight the war?" And I felt it was up to me. I was older, I was thirty-eight. It was up to us older people to stop the war.

So the question was, what could I do to help end the war if I were willing to go to prison? Once I thought about it, it wasn't that hard to decide. As for my readiness to go to prison, I'd already risked my life *for* the war when I had gone to Vietnam. And if I could do that, I could now go to prison if necessary, as the price of opposing it.

I realized that one thing I could do was to copy the study I had on the origins of the war. Now it wasn't immediately obvious that this was a useful thing to do, because the study had ended with 1968. We had a new President, and the study was now history. And it wasn't clear that pointing out the history of the war would help Nixon end it. If I'd had documents to prove what I'd been told about Nixon's plans to expand the war, I would have put those out instead; that would have been much more relevant. But I didn't have such material. I had a study that I had done for Kissinger and Nixon earlier that year, suggesting their interest in expanding the war, but it didn't really prove anything. So that wasn't going to do it, either.

I knew that no President would pull out of the war if he felt that it was his to lose, if it was in any way associated with him. If he did, he'd be the first President to lose a war, and no one wanted to live with that distinction. He'd rather keep going, no matter how many people died, to save face. But if Nixon could be induced to get out before he was too strongly associated with the war, he might be willing to do it. And I thought the only way he would do this was to get the Democrats to share the responsibility with him, to say, "Mr. Nixon, this isn't your war. This is our war. We got ourselves in. But don't make the same mistakes. Get us out, and we'll stand with you. We'll take the responsibility." I actually called up several key Democrats: Paul Warnke, who had been assistant secretary of defense, and Harry McPherson, who had been the assistant to President Johnson. Both of them were in a high-level policy group called the Democratic Policy Advisory Group. But they weren't willing even to try.

The Democrats weren't willing to do any such thing. They said the Democratic Party would suffer, they'd be accused of stabbing Nixon in the back. To lose a war after getting us into it in the first place would have been disastrous for the Democratic Party. One of them said there would be a bloodbath "such as you and I have never seen" – meaning a political bloodbath, of course, like McCarthyism. He said, "And that means you and me, Daniel. We'd get it." I said, "Well, you might be right. That might very well happen. But I'm not willing to see the real bloodbath go on in Vietnam, to see more Americans being killed in order to save the Democratic Party or to save myself politically." So they couldn't rise to that, and I began to realize that if anyone was going to say anything, it would have to be me.

I had already been helping draft a letter from Rand to the *New York Times,* which identified ourselves as being at Rand and said, "We should get out of Vietnam in one year." Six of us at Rand were involved in that, and it was a very controversial thing to do. Later they clung to their jobs by their nails – their bosses tried to fire them for writing this very unusual letter. I had also written to other people, trying to get us out, and it all looked more promising than the Pentagon Papers – which, as I said, was just past history.

But then something else occurred, which led to my final decision. On September 30, 1969, while I was in the midst of all this activity, I was on the beach in Malibu. I was reading the paper, and the story that day was about a special-forces murder trial that had ended and had been dropped. Eight men associated with the special forces in Vietnam, including the commander of the special forces in Vietnam, a colonel, were on trial for having killed a Vietnamese agent, one of their own, whom they suspected had also been working for the Vietcong. They had taken him out over the South China Sea and shot him with a .22, with a silencer, and then dumped him over the side, weighed down with chains. His body was never found.

Well, it was a very controversial case. Most Americans complained, "Why are we persecuting our special forces for killing a Vietnamese agent? What the hell! This is war. And why are we putting these brave men on trial for doing what they thought was right, after their bosses decided that it was the right thing to do?" That was the main feeling about it. It had been a very unpopular trial, as was Calley's trial for My Lai later: the most popular single act that Nixon ever took was to shorten Calley's sentence and put him under house arrest, out of the stockade. Anyway, these men were then taken out of confinement because there was such a popular outcry. That had all happened before September 30, but it had all been going forward to trial.

Now the trial had been dropped, and the newspaper stories on it had a peculiar form which I'll just summarize this way: "The Secretary of the Army has dropped charges, saying that it's his decision to do so." Clearly, this was a lie. It was very clearly a White House decision. The White House was saying that it had not taken part in the decision, but that was obviously a lie. And the Secretary of the Army was saying that he did this because the CIA wouldn't testify, but that was obviously a lie, too, because the CIA could have been ordered to testify by the President. In fact, Haldeman's diary later revealed that indeed the White House had ordered the CIA to say that they could not testify.

But the lies didn't end there. General Westmoreland's successor in Vietnam, General Abrams, said that he had no choice but to bring charges, and now he was being overruled by the Secretary; it was a straightforward murder case, he had to bring charges – he had no choice. But that was obviously a lie. There were murders every day in Vietnam, on a massive scale and on a small scale. They didn't have to bring charges. They had never brought charges before. Contrary to what was said, Abrams had brought charges because he'd been lied to by the colonel in charge of the special forces – lied to about the circumstances under which this agent had been killed. And the colonel had ordered all of the people under him to lie, too. So Abrams was very angry at being lied to by all of these captains, majors, sergeants, and warrant officers, but especially by the colonel, who was a West Point man. That is why he had brought charges of murder against them.

I lay there and I thought: this is the system I've been part of for fifteen years, including the Marine Corps. It's a system that lies automatically at every level of command, from the sergeant to the commander in chief, up through the captains, the majors, the colonels, and the generals. And the Secretary of the Army. To conceal a murder. And I said, I'm not going to be part of this system any more. I'm not going to be part of this lying and this murder and this war.

At that moment, I had 7000 pages of top-secret documents in my safe, evidence of a quarter century of lying and murder, aggression, broken treaties, deceptions, stolen elections – I had proof of it all. It wasn't clear how much good it would do to bring it all out. It might not help, but it couldn't hurt. And at any rate, I was going to get out of this system of lying.

The lying didn't bother me so much as what we were lying about. After all, if lying bothers you too much, you can't be in the government for a month, and I'd been in for fifteen years. The Marines didn't involve me in too much lying that I can remember, but every month in the executive branch did. It was what the lying was about: lying about mass murder. To continue the war with no hope of any benefit to any American interest was murder; it was unjustified homicide. Even if you could justify some killing, you couldn't justify *this* killing.

I figured that what I was going to do would put me in prison for life. They would send me to prison forever, and I would not have done that just to set a lie straight. But the point was that there were many more murders coming; it was the future prospect of more lying and more murder that concerned me.

So I decided to phone a friend of mine who had been fired from Rand for exposing serious crimes in Vietnam. His name was Tony Russo; he had become an embarrassment for Rand, and so they had chosen to get rid of him. And I asked him if he knew where there was a Xerox machine. He said he had a girlfriend who ran an ad agency. She had a Xerox machine, and we could use it if we paid for it. So we started that night, copying the Pentagon Papers.

Even before I went, as I was getting dressed that night, I thought of two principles in the light of what I'd now decided to do. One was that no one would ever again tell me that I had to lie. I could imagine lying again in my life, but I said, "I will take the responsibility for it. I won't tell myself or anybody else that I'm doing it because I was ordered to lie. I'll be responsible if I feel it's necessary."

And the other principle concerned the question of violence. I was wrestling then, as I have been ever since – thirty years, almost – with the question of whether it's ever justifiable to kill someone. I had even talked to Pastor Niemöller at the War Resisters League conference. He'd been a decorated U-boat commander in World War I but was later imprisoned, as the bishop and pastor of Berlin, when he opposed Hitler's treatment of the Jews.

So I thought,"No one will ever again tell me that I have to go to war to kill someone." Now they always say that in war killing is justified – you share the responsibility, and they say,"It's our responsibility. It's not your responsibility." But none of that made any sense. You cannot delegate such a responsibility to anybody else. You can't delegate your conscience to somebody else who says it's all right to kill somebody. If you kill somebody, you have to take full responsibility. So that answered the question about the draft. As Dan Berrigan once said,"To register for the draft is to announce your availability to kill on command."

I came to believe that the draft represents a system based on violence, on the legitimacy of organized violence, that gives far too much power to national leaders and to national policy. It really is worthy and important to announce, if you're capable of doing it, that you're not available, that you don't accept the authority of such commands. I know it used to be true, and I suspect it still is, that the constitutionality of the draft as involuntary servitude has never been tested in the Supreme Court. And there's a very good reason for that: the constitutional basis for it would be very, very suspect.

At any rate, the bombing continued. And the bombs did not drop themselves. They were dropped by humans who planned the missions, who loaded the bombs, who flew the airplanes, and who knew just as well as everybody else in the world that they had no justifiable purpose whatever. But it went on – for five more years.

Of course, my own experience had shown me that it was very foolish to think that these higher-ups, with their "better information" and their "higher responsibility," their "judgment" and their "knowledge" and their "values," could ever be believed. Look at the examples: the missile gap, which didn't exist; and going back earlier, the Manhattan Project, which recruited people to deter a German risk that didn't exist.

You could delegate the responsibility for this killing to a national leader only if you had managed to remain ignorant of all this history. To say that these people should be given the benefit of the doubt implied either that you were too young to know better, or that you were some kind of an idiot who was not aware of what had been going on for the last twenty years. And how often these people had made terrible and stupid judgments in the face of all the information! People, men in power, are the last people in the world whose judgment you should trust with these decisions. You're much better off trusting their wives or their daughters, but unfortunately they never get near the decision.

The Pentagon Papers came out in 1971. I had given them to the Senate in '69, but the Senate didn't hold hearings. They were afraid of the consequences. So two more invasions took place, Cambodia and Laos. My son turned eighteen, and the war was still going on. I was put on trial, facing 115 years in prison (my feeling that they would try to put me away forever was not off the mark). Twelve felony counts. The trial ended; Nixon was kicked out of office. And the war was still going on in Vietnam. U.S. troops were no longer directly involved, but we were still paying the entire cost of the bombs; it was still our war. We were still paying the Army, paying for the bombs and the planes. We had equipped South Vietnam with the third-largest Air Force in the world, after Russia. So it was not exactly their war. It was still very much ours.

Well, I didn't pay the price that they had in mind for me. Nixon was much more interested in destroying me outside of a trial than inside. The newly released transcripts of the Nixon tapes are very funny, actually, on the importance of getting Ellsberg and destroying him, not in the courtroom but in the press, by reputation, smearing me, and so forth. They broke into my former psychoanalyst's office in Los Angeles, not so much to smear me as to blackmail me, to get information which they hoped would buy my silence. Well, none of that happened. They didn't find too much, and what they found they couldn't use. They didn't manage to hurt me very much. They wanted to, but they failed.

The actual price I paid, which was quite significant, was that they made me into a kind of Typhoid Mary – a contagious, dangerous person from the point of view of my professional life. All my friends had clearances, and their clearance status and their access to office and to information would be jeopardized if they dealt with me at all. Seymour Hersh, the newspaper man, is one of my oldest friends, because he goes back to the period when I was on trial. But I don't have very many friends that go back before him, because I lost them all. I essentially lost all of my friends. Of course I had some relatives and a few old school friends that I see once in a while, but by and large the people from my adult working life have all vanished. It was as if I'd emigrated, and it was a very stiff price to pay. But I don't regret what I did.

Looking back, my decision to put out those documents had no immediate effect. It affected people's minds, but the war went on. And I hadn't risked my life just to affect people's minds. I wanted to end the war, and the war didn't end for another four years. But still it seemed very worthwhile at the time, and in the end it did turn out to be very effective.

The reason for this was not apparent back then. Nixon was carrying on his secret policy of expanding the war, which would have been very unpopular if known. And I was a threat to those secret plans. He feared that others who knew of his secret policies might imitate me, but even more he feared that I had documents proving his actual policy. I didn't. I had a couple, but not enough; he thought I had far more evidence than I ever had. So he had very strong reasons to try to shut me up, and that's what he tried to do. He ordered a bunch of people to "incapacitate" me, to kill me or at least to beat me up, at a critical moment in 1972. He tried to blackmail me so that I wouldn't put out the documents he feared I had. Blackmail, of course, is a crime. And so it had to be covered up.

The people who broke into my doctor's office and who later tried to beat me up on the steps of the Capitol were, by an error of Nixon's, the very same people he sent into the Watergate Hotel. If they hadn't been, if it had been a different set of people, the war might have gone for several more years. But the moment these people were caught in the Watergate, Nixon worried that they would tell about their earlier crimes against me. And so he bribed them into silence. That was another crime, and it, too, had to be kept silent. More and more people knew; they all had to be paid off, and sooner or later some of them began to talk.

To this day, the actual break-in at the Watergate Hotel cannot be traced to Nixon. In fact, Nixon says throughout the tape transcripts, "I didn't know they were going into the Watergate." That's probably a lie, but we don't have any proof. So he would never have been forced to resign because of Watergate alone, or because of the campaign contributions or the other dirty tricks. The only things that could be traced to him were the things he did against me and other antiwar activists, the cover-up of his direct and personal involvement in trying to destroy me. If it had just been Hunt and Libby and those guys in the Watergate, he wouldn't have had to cover things up. Let them go to jail. The cover-up was to protect himself.

In the end he got caught and he had to resign from office, which allowed Congress to end the war. Congress could have cut off the money when he was still in office, but he would have defied them, which Ford wasn't willing to do. So Ford, facing a congressional cutoff of money, had to let the war end. But the bombing could have gone on many more years. It might have gone on another ten years.

VIETNAM VETERAN STEVE BENTLEY has fought long and hard to reveal the way in which thousands of Vietnam veterans were exposed to a toxic chemical defoliant, Agent Orange. The former chair of the Vietnam Veterans of America PTSD and Substance Abuse Committee has worked tirelessly to pierce the barrier of silence that surrounded the Pentagon's use of such chemicals for decades. Steve writes:

Nationally, 220,000 Vietnam veterans have requested physical examinations from the Veterans Administration because of suspicion that their health problems are caused by exposure to the 12 million gallons of herbicide that was sprayed on Vietnam…

[Recently, chemical companies agreed to a] $180 million settlement… in order to keep [them] from going to court…In order to receive any part of that money, the veteran must prove he was exposed to Agent Orange in Vietnam, has been suffering from long-term disability, and is 100-percent disabled by Social Security standards. This last requirement obviously leaves out the vast majority of Vietnam vets who work despite health problems, and it leaves out all those whose children have suffered birth defects and cancer.

The chemical companies seem to be saying that, while they don't really believe these problems exist at all, it is absolutely necessary for us to be totally disabled by these non-existent problems in order to be compensated even a little bit. If you meet these requirements, the average expected payment is $5700. That is, unless you were unfortunate enough to have these imaginary problems kill you; then the maximum death benefit is $3400. To those families most affected, even inadequate compensation is better than nothing. However, the attitude this settlement reinforces in many Vietnam vets is summed up in the bumper sticker "Sprayed and Betrayed"…

Steve goes on to note that the very government which sent him to Vietnam has concluded, via its Centers for Disease Control, that it will not bother even to look into such health problems because they are "statistically insignificant."

Ironically, these statistically insignificant soldiers who were heavily exposed to dioxin also turn out to be those who were most involved in combat by virtue of the fact we were the ones in the jungle. This fact makes further mockery of the VA's much-touted slogan, "To care for those who have borne the battle."[5]

Steve is incensed by a government that "panders to veterans about the symbols of justice while real live human beings stumble from one alley to another." Spurred by the hype following Operation Desert Storm in 1991, he put his thoughts to paper in an article entitled "In the Name of Freedom":

In the name of freedom for the people, the leaders of the U.S. government spent $170 billion and 60,000 American lives in Vietnam. Now, 20 years later, the administration continues to drag out the MIA/POW issue, thereby blocking diplomatic relations with Vietnam. The ultimate result is continued economic devastation for Vietnam and ongoing untold suffering for the very people we professed to care about.

Again, in the name of freedom for the people, the U.S. leadership spent well over $500 million on many forms of lethal aid in hopes of facilitating the overthrow of Nicaragua's Sandinista government. When the people of Nicaragua voted out the Sandinistas, the hard-liners were quick to take credit and justify the expenditure in the name of freedom for the people.

Now, after years of war and U.S.-backed embargo, most Nicaraguans live in such conditions of squalor that a cholera epidemic threatens to kill tens of thousands of them. It's estimated that $20 million of medicines would quell this epidemic. The United States is able to come up with less than $2 million for the people.

In the name of freedom for the people, the United States government helped to create and nourish a madman in Iraq and then, when he acted as madmen do, our leaders spent billions of dollars and hundreds of American lives blowing at least 100,000 Iraqi people to smithereens. Meanwhile, we left the madman in power.

Recently, a Harvard team of health experts returned from Iraq to inform us that 200,000 Iraqi children are destined to die if we continue sanctions. We're told, however, that we must continue sanctions so we can get the madman out of power in the name of freedom for the people.

In the midst of all this, President Bush wants to give "most favored nation status" to a bunch of power-hungry old men who slaughtered their children in the town square.*

Who's kidding whom? This government doesn't give a damn about freedom for the people. It has nothing to do with freedom. It's about power, it's about money, it's about racism and elitism and greed. It's about pride and stupidity and ugly, insidious forms of paranoia, projection, and delusion, but it has nothing to do with a commitment to freedom for the people [and] even less to do with their well-being.

To make matters worse, it appears the majority of the citizens that make up this nation buy into this nationalistic babble like so many freshly caught carp. Which all serves to remind me of the words of Thomas Jefferson, "I tremble for my country when I reflect that God is just."[6]

ONE CANNOT SPEAK of war and its attendant dishonesty without thinking of the massacre at My Lai, recounted earlier; its systematic cover-up remains one of the greatest fiascoes in modern history. Although the Army knew about the murderous rampage almost immediately, news of the "incident," as it was quickly renamed, appeared in the American press only two years later. As angry journalists mobbed the Pentagon, the Army's general counsel, Robert Jordan, told them:

Some of you enterprising members of the press have been poking and digging around the last few days and you have identified some people

*Reference to the June 4, 1989, massacre at Tiananmen Square, Beijing, China.

that are identified in this and you've found out where they are and you've gone out and interviewed them. We don't consider it within our purview to order everybody who is in the Army not to talk to you...[7]

Look, let's be realistic about this. Enough information has come out about this case...that any question about the Army protecting itself is really kind of silly, you know, the fat is in the fire. If you're talking about trying to cover up, we aren't trying to cover up and we haven't been trying to cover up.[8]

But Jordan's defensive assurances were at odds with Army policy that had been in effect ever since the beginning of the war. This policy had been underscored in October 1967, as General Westmoreland sought to stem the flow of increasingly harrowing reports and photographs to the American media.

In a confidential letter to his deputy, General Bruce Palmer, Westmoreland had called for restraint, warning that "the war in Vietnam is being degraded by a growing segment of the press and public at large." He demanded that any activity "reflecting unfavorably on the military image" be eliminated, and specifically listed prostitution, excessive drinking, and violations of the Geneva Conventions and the Law of Land Warfare. In a closing line that lays bare the true underpinnings of the democratic process, he reminded Palmer that the "tensions of the forthcoming election year will focus to a major extent on virtually every aspect of our effort here."[9]

When the internal Army investigation into the My Lai "incident" was disbanded on April 8, 1968, only three weeks after the killings, Brigade Commander Henderson congratulated the men of "Charlie Company" himself: "The praiseworthy role

of units of the 11th Infantry Brigade directly reflects your expert guidance, leadership, and devotion to duty…The quick response and professionalism displayed during this action has again enhanced the Brigade's image in the eyes of higher commands."[10] The message to the murderers was clear. There was no need to worry about further investigations, and the truth would be quietly buried.

The Americal Division's commander, Major General Samuel Koster, was forced to resign his newly appointed position as superintendent of West Point when it was found that he had deliberately withheld reports indicating that a massacre was in progress. He was hardly repentant. From the balcony overlooking the dining hall, he announced to the academy's cadets that he was "requesting reassignment to save the academy from further publicity." He spoke of the cherished principles of West Point ("duty, honor, and country") as the guide for his military career and added, emotionally, "I shall continue to follow these principles as long as I live. Every one of you can have a wonderful experience in the military profession. To serve one's country true and faithfully is one of the highest callings…Don't let the bastards grind you down!"[11] Koster never identified who the "bastards" were.

As the truth about My Lai unfolded, a newly promoted Alexander Haig kept the White House appraised of developments in the case. Secretary of State Henry Kissinger, in turn, passed the information to White House Chief of Staff Haldeman and President Nixon. "Now the cat is out of the bag," he wrote to the Chief of Staff only days before Lt. Calley was formally

charged with a capital offense."I recommend keeping the President and the White House out of the matter entirely."[12]

An exasperated Richard Nixon was less delicate, and as public reaction to the news became more hostile, he blamed the "dirty rotten Jews from New York" – ostensibly, the editorial management of the *Times.* When college students scheduled a nation-wide antiwar protest for October 15, he turned angrily to Chief of Staff Bob Haldeman: "It is absolutely essential that we react insurmountably and powerfully to blunt this attack." He was talking not of the NVA, but of American citizens, marching unarmed through ivied campuses.[13]

British investigators Bilton and Sim provide a rare view into the world of the Vietnam-era White House, and their portrayal of deceit at the highest levels of government lends new weight to the concept of media control:

Nixon's tactic, true to form, was to go on the attack, to turn the stories about My Lai on their head...At a meeting with Haldeman on December 1, the President secretly ordered that the campaign against media bias be kept boiling and that a covert offensive should be mounted on My Lai. Haldeman's personal notes of the meeting, which remain among Nixon's White House files in the National Archives, reveal the President's bidding: "Dirty tricks – not too high a level; discredit one witness, get out facts on Hue; admin line – may have to use a senator or two, so don't go off in different directions; keep working on the problem." Nixon condemned the atrocity in public. But one night, according to a White House assistant, Alexander Butterfield, he spent two hours attacking the press for the My Lai publicity.[14]

Daniel Patrick Moynihan, then counselor to the President, wrote a personal memo to Nixon, expressing his fears:

It is clear that something hideous happened at My Lai...I would doubt the war effort can now be the same, nor the position of the military...I fear the answer of too many Americans will simply be that this is a hideous, corrupt society. I fear and dread what this will do to our society unless we try to understand it...I think it would be a grave error for the Presidency to be silent while the Army and the press pass judgment. For it is America that is being judged.[15]

The memo fell on deaf ears, and as time would tell, Moynihan was only partly right. The antiwar movement was certainly galvanized, but the position of the military remained, predictably, unchanged. Public outcry gave way to the myth of moral superiority; like famine and the plague, war crimes were the property of uncivilized foreigners.[16]

The cover-up of My Lai continues to this day. Until 1998 few had heard the story of Hugh Thompson, a helicopter pilot who accidentally came upon the massacre, in progress, as he flew over the jungle on a scouting mission. Thompson set his chopper down between a group of frightened civilians and the advancing American soldiers, and threatened to open fire on his own men should they kill any more civilians. He managed to rescue a three-year-old boy and ten other Vietnamese who had hidden in a bunker. At the time, the Pentagon's official report hailed Thompson as a hero; the Army offered him a Distinguished Flying Cross in 1969.[17]

More recently, in 1996, Thompson's name came up again – he was recommended for the prestigious Soldier's Medal, for

bravery. But an internal Pentagon memo from an anonymous major advised Army Secretary Sara Lister to hold off: "We would be putting an ugly, controversial, and horrible story on the media's table," it said. "...I recommend sitting on this until clear of the election." In November 1997, the Army tried to play down Thompson's role once again: this time, they suggested giving Thompson the medal in a "private ceremony."[18] In March 1998, as publicity surrounding the thirtieth anniversary of My Lai built, he finally received the prestigious Soldier's Medal in a ceremony at the Vietnam Veterans Memorial in Washington, DC.

THE 1937 RAPE OF NANKING dwarfs the My Lai massacre in sheer numbers, if not in cruelty. Despite its magnitude, however, it remained largely hidden in the annals of World War II history until recently.

The broad details of the rape are, except among the Japanese, not in dispute. In November 1937, after their successful invasion of Shanghai, the Japanese launched a massive attack on the newly established capital of the Republic of China. When the city fell on December 13, 1937, Japanese soldiers began an orgy of cruelty seldom if ever matched in world history. Tens of thousands of young men were rounded up and herded to the outer areas of the city, where they were mowed down by machine guns, used for bayonet practice, or soaked with gasoline and burned alive. By the end of the massacre an estimated 260,000 to 350,000 Chinese had been killed. Between 20,000 and 80,000 Chinese women were raped – and many soldiers went beyond rape to disembowel women, slice off their breasts, nail them alive to walls.

Why has this abomination remained so obscure? The number of deaths far exceeds that of Hiroshima and Nagasaki combined, and yet the horror of Nanking remains virtually unknown outside of Asia.

The Rape of Nanking did not penetrate the world consciousness in the same manner as the Jewish Holocaust or Hiroshima, because the victims themselves remained silent. The custodian of the curtain of silence was politics. The People's Republic of China, Taiwan, and even the United States, all contributed to the historical neglect of this event for reasons deeply rooted in the Cold War. After the 1949 Communist revolution in China, neither the People's Republic of China nor Taiwan demanded wartime reparations from Japan (as Israel had from Germany), because the two governments were competing for Japanese trade and political recognition. And even the United States, faced with the threat of communism in the Soviet Union and mainland China, sought to ensure the friendship and loyalty of its former enemy Japan. In this manner, Cold War tensions permitted Japan to escape much of the intense critical examination that its wartime ally was forced to undergo.[19]

The cover-up of Nanking is typical of the selective silence and propaganda of the "Good War." And it is by no means the only incident that was sidelined willfully. Recently, *U.S. News and World Report* admitted that the righteous Allies themselves were hardly immune to Hitler's greed:

Europe's past keeps sending awful bonbons to its present. In recent months, we have learned that Swiss banks piled up millions of dollars in assets stolen from Jewish victims of the Nazis. Portugal, Turkey, Spain, and Sweden also traded in Nazi loot. The French, who pride themselves on their resistance movement, in reality produced more collaborators –

like Maurice Papon, who is on trial for his role in deporting Jews in World War II – than heroes. And now we discover that the United States and Britain knowingly melted the belongings of Holocaust victims into gold bars and distributed them to central banks in Europe as reparations. Everyone has been implicated in a crime once attributed to a single nation: Germany… [20]

Philip Caputo, a war correspondent who later became a best-selling novelist, supported the Gulf War but complained nonetheless about what he perceived to be excessive control of the media. Clearly, the military had learned well the lessons of Vietnam, at least as far as "bad press" was concerned: "The Pentagon…censor[ed] the press beyond reasonable need," he wrote. "We see only what it wants us to see, which is a war weirdly sanitized of the pain, fear, and death that are the essence of war."[21]

Daryll Byrne, a medic in the ground war, went on to speak about the control of information during Desert Storm and Desert Shield:

We've talked about the press and how they deal with war. In the Persian Gulf, at least where I was stationed, there were no TV cameras, no reporters. So the picture was painted by people who weren't there. What you saw back here was mostly military footage.

When I came back, I was amazed at the videos, the television programs, the magazines with the detailed accounts of everything that had occurred. I learned more about the Persian Gulf War after I came back than when I was there fighting it. When you're at war, you don't get this information. What you get comes from the military, and that information is always tainted.

Retired Army Colonel David Hackworth, the nation's most-decorated living veteran, covered the Gulf War for *Newsweek*. He, too, was displeased and berated his former colleagues for their "paranoia and their thought-police who control the press":

Although I managed to go out on my own, we didn't have the freedom of movement to make an independent assessment of what the military is all about. Everything was spoon-fed. We were like animals in a zoo, and the press officers were the zookeepers who threw us a piece of meat occasionally...I had more guns pointed at me by Americans and Saudis who were into controlling the press than in all my years of actual combat.[22]

Of course, there were plenty of things the Pentagon needed to hide. Not the least of these was the secret exposure of thousands of American troops to new and highly radioactive artillery which was used in the Gulf War for the first time. An extremely dense by-product of the higher-grade material used in nuclear bombs, depleted uranium (DU), soon caused the Gulf War veterans to fall ill by the thousands. Besides being exposed to DU, Gulf War soldiers were given experimental vaccines and drugs to counteract a variety of toxic agents developed for use in chemical warfare.

Paul Sullivan, director of the National Gulf War Resource Center, himself suffering from Gulf War illnesses, offered his perspective in an interview in September 1997:

For the military to carry out its objective of killing people, it has to have secrets. It has to have secrets about its weapons, secrets about its plans, secrets about its training. And once you have a secret, then you start to lie, because you have to deny the existence of certain things. When people find out about that, then you have to lie again.

We're seeing this very clearly with Gulf War illnesses. The secretary of Defense, the secretary of the Department of Veterans Affairs, and the secretary of the Department of Health and Human Services all lied in testimony in front of Congress and again in letters sent to the veterans. They said that they had conducted a thorough investigation and found that no one, at no time, was ever exposed to any chemicals during the Gulf War. Well, through painstaking research over several years, Gulf War veterans discovered the web of lies. We've found soldiers who received Purple Hearts for being exposed to chemicals during the Gulf War, and in order for them to be exposed, chemicals had to be present, and they had to be released. There is no other explanation. The whole web of lies comes crashing down.

We've also discovered that the Pentagon knew before the war that the chemical detection and protection equipment they gave us was no good. They knew it, and they knowingly sent people into battle without the protection they needed in order to survive. Any reasonable person will conclude that the Pentagon intentionally sacrificed Gulf War veterans, knowingly sending them into a chemical battlefield to die.

Approximately 155,000 veterans of the Gulf War are sick. And the Department of Veterans Affairs refuses to release the number of dead. That, in and of itself, heightens speculation, based on this history of lies. The Department of Veterans Affairs and the Department of Defense are still withholding their information on casualty count. So the war continues.

The Pentagon is not talking; it's denying that anyone was ever exposed. But that's a big lie. They've already admitted (although they had to be dragged kicking and screaming) that during the Reagan and Bush administration, American chemical companies sent all manner of biological and chemical warfare agents, along with "delivery systems," to Iraq.

Paul's assessment is substantiated by thousands of other incidents, including the following letter submitted to the Department of Veterans Affairs by former Army doctor Asaf Durakovic:

Dear President Clinton: February 11, 1997

I am bringing to your attention the conspiracy against the Veterans of the United States.

In the Persian Gulf War some veterans were exposed to radioactive contamination with Depleted Uranium. I personally served in the Operation Desert Shield as a Unit Commander of 531 Army Medical Detachment. After the war I was in charge of Nuclear Medicine Service at Department of Veterans Affairs Medical Center in Wilmington, Delaware. A group of uranium contaminated U.S. Veterans were referred to my attention as an expert in nuclear contamination. I properly referred them for the diagnostic tests to different Institutions dealing with transuranium elements. All of the records have been lost in this Hospital and in referring Institutions. Only a small part of information was recorded in Presidential Advisory Committee report on Gulf War Illnesses. Recently I received an order by the Chief of Staff of this Institution to start the veterans examinations again since all of the records have been lost.

Today I was informed in writing that my job was terminated as a reduction in force. I have been at this position for over eight years with an outstanding job performance and I am convinced with certainty that my elimination from the job is a direct result of my involvement in the management of Gulf War Veterans and discrimination for raising nuclear safety issues.

The lost records, lost laboratory specimens and retaliations which are well documented point to no less than conspiracy to terminate my efforts of proper management of Gulf War Veterans. I am sure that you will have an interest in this matter for the benefit of the veterans of The United States of America.

Most respectfully
(signed)
Asaf Durakovic, M.D., D.V.M., M.Sc., Ph.D., F.A.C.P
Professor of Radiology and Nuclear Medicine
Chief, Nuclear Medicine Service, VAMC Wilmington
Colonel, U.S. Army Medical Corps (R)

OF ALL THE EVIDENCE of the "hush" enveloping the Gulf War, I find veteran Carol Picou's story most damning. Her saga of ongoing suffering is but a microcosm of the selective silence that surrounds every war; to hear and see her pain firsthand gives immediacy to a problem otherwise buried by statistics.

Carol speaks out with all her remaining strength, but there are many thousands like her who suffer quietly, waiting for someone, somehow, to shed light on the "misunderstanding." Carol knows better than to hope for apologies or recompense, but she isn't ready to die either.

When we arrived in the Gulf, we were given pills, and we were ordered to take them three times a day, every eight hours. They told us it was to protect us in case of chemical warfare. So we took them, and within one hour my muscles started to ache and my eyes began to tear. Very soon my nose was running, and I was drooling and my muscles were twitching.

"Guys," I said, "there's something wrong." But I was ignored. They told us to keep taking them, and each time I did, I had the same symptoms. Finally on the third day I secretly spit my dose into a Pepsi can. And the symptoms disappeared...

Later, at a field hospital, we were told to stand in formation, and we had to take the pills again. By then many of the other soldiers were also complaining of muscle aches and cramps. I refused to take the pills, and when the platoon sergeant noticed, he pressured me, saying that I needed to take the pills "for my own protection."

One hour later, my symptoms were back. I talked to the NBC officer, the nuclear-biological-chemical doctor. "Well," he said, "You just proved that it's peaking and that it's working on your nervous system. Just keep taking it."

We were forced to take this drug for fifteen days, and by the time we finally got word to stop, the war was over. And the damage was already done.

By the time I came back to the United States I was having more problems. I wasn't sweating anymore, I was hoarse, and I was having urinary difficulties. The Army reassured me: it was a change in my diet. I was back in the United States now, and all of this would straighten itself out. But it never did.

Within a month of my return in July 1991 I was incontinent, and I had diarrhea. Then I started to lose my memory – I'd forget to pick up my son after work. In October I had an accident in the doctor's office, and they finally sent me to urology. By January 1992 I was wearing diapers. I wanted to know why. Why was I getting like this? Why was I having these problems? Nobody could answer my questions, so I went public: what's wrong with us soldiers?

It took off from there. I started to do research with a veteran who thought it was depleted uranium, and a researcher at the Library of Congress helped me to write a paper; I was called to testify before Congress the next spring.

I was actually the first active-duty soldier to testify in June of '93. But I didn't go in military uniform. I went on my own, because I didn't want any repercussions. I testified before Congress, and of course no one believed anything. The government said that all of this didn't happen...

Then I started getting phone calls: better be prepared for what's going to happen to you. Vietnam veterans would call and tell me I had no idea of what I was getting into. I got one phone call, and the caller was holding something by his voice to distort the sound: you don't know me, but they're watching you. They know what you're up to. And then he hung up.

The Army also threatened to destroy my career, and they told me I was going to lose my private health insurance and my life insurance. That happened. They discharged me, non-combat-related, with a $20,000 medical bill, and I was really stuck. The insurance people said it was combat-related, and the Army said it was not. And then my insurance company dropped me anyway. My husband lost his job fighting them...

Then we tried the VA hospital, and they said I should be able to get Social Security. But the people at Social Security said I was too young and too educated. I wasn't eligible. Of course I wasn't eligible for unemployment, either. They just pass the buck from one person to another.

My illness has destroyed my life. My son is afraid I'm going to die, because my friends are dying... We used to play ball and ride our bikes together, but I don't anymore, because I get too tired. I can't even go watch him play soccer anymore. I have no feeling from my waist down, and so I soil myself. Then my little boy says, "Mommy, we need to go home." He's been through a lot.

I obtained a document in 1994 that described everything we soldiers had been exposed to: weapons, insecticides, pesticides, anthrax, experimental vaccines, pills. Shortly after receiving the report, we were in conference, on the phone, with a whole network of ex-service people, uranium miners, and Native Americans, talking about what we could do to get this exposed. My husband read the report over the phone and said, "We'll get it in the mail to you on Monday."

I left the house with that report later the same day, and then forgot to bring the papers in from the car. And at 1:45 the next morning I awoke from a sound sleep. Something said, "Carol, wake up and look out the window." I did, and our car was totally engulfed in flames.

I've had the opportunity to go back into Iraq since the war; I spent nine days in the area where we fought. People came from the surrounding villages and brought me their children. I held a three-year-old child who was no bigger than my arm. I held babies that were lifeless. Then I was in Baghdad for three days, in and out of hospitals, meeting people who are affected by depleted uranium. Their babies and their children are dying; their soldiers are dying of strange cancers. That was very hard to see.

At first, I didn't hug those Iraqi children. I didn't embrace them like I would have liked to, because I felt responsible. I had helped to destroy these people. But at the same time I felt I had a responsibility to expose their suffering, to try and help these people.

In the United States we have soldiers dying, too. The brain tumors and the cancers are just amazing: young soldiers, twenty-three, twenty-four years old, with non-Hodgkin's lymphoma. Our babies are being born without arms or legs, missing eyes, missing ears, hearts on the wrong side.

But the Iraqi soldiers and children have identical problems, and they can't even get medicine. They don't have the means to provide even basic supportive care. And I ask myself, "Which is the more barbaric nation?"

An Iraqi colonel I met has the same symptoms I do: the neurological damage, the suppressed immune system, the bone and joint deterioration. He said that out of the twenty-six years that he has served in the Iraqi Army he's never seen a weapon so deadly. Certainly it was a different kind of war, unlike anything anyone's ever seen: the tanks were burned, and all the bodies in and around them were charred, or even melted onto their sides.

Their land is permanently contaminated. When I went back we took Geiger counters, and we climbed into some of the same vehicles that we had destroyed. They were highly radioactive, six years after the war. Even animals are being born deformed. We have destroyed their land, their people, their future.

This is where I struggle, because when I joined the military I thought I was just serving my country. At one time I felt proud to be an American – I felt it patriotic to join the military. But the military teaches you to "search and destroy." It is not there to help, to build back up…

It took the Pentagon twenty-two years to listen to the problems of Vietnam veterans, forty-six years to admit to the mustard-agent problems of World War II. And we're in our sixth year with the Persian Gulf. So we just have to keep at it.

LITTLE DID LEONARD DIETZ know, when he came home a scarred veteran of World War II, that he was to become an advocate for veterans of another war nearly fifty years into the future. That chapter of his life would begin when "new" veterans like Carol Picou were still in grade school, learning about democracy and freedom and the "Good War." And, once again, it would be a chapter cloaked in propaganda, secrecy, and deception.

In autumn 1979, Leonard worked as a nuclear physicist at the Knolls Atomic Power Laboratory in Schenectady, New York, which was operated by the General Electric Company for the Department of Energy. While trouble-shooting a problem for GE's radiology group, he and his colleagues accidentally discovered depleted uranium aerosols in environmental air filters at the Knolls site. They traced the source of the uranium contamination to the National Lead Industries plant in Colonie, 10 miles east of the Knolls site, near Albany. At the time, National Lead was fabricating depleted uranium penetrators for 30-mm cannon rounds. Leonard's team also discovered depleted uranium at another site 26 miles north of the National Lead plant, suggesting that the radioactive aerosols were scattered over an area of 2000 square miles.

The plant was subsequently shut down by state health officials, but for an unrelated incident: National Lead was exceeding the monthly DEC radioactivity limit, equivalent to only one and a half of the penetrators used in every 30-mm round.

More than a decade later, when word of the military's use of depleted uranium penetrators in the Gulf War began to leak out, Leonard was deeply alarmed. As a physicist, he knew why such deadly material was being used. Uranium metal, when alloyed with titanium, is extremely dense and hard, and the penetrators, which are thin rods of this alloy about two feet long, are pyrophoric. That is, they combust upon impact: Traveling at nearly a mile per second when they strike a tank, they ignite and burn, literally melting through otherwise impenetrable tank armor. The combustion also releases a deadly cloud of aerosolized uranium particles.

This airborne uranium can be taken into the lungs, where the tiny particles enter the bloodstream and migrate to the lymphatic system and virtually everywhere else in the body. And there they stay. Radioactive contamination from depleted uranium aerosols is permanent, and cells surrounding the particles are irradiated for a lifetime, causing cancers, birth defects, and other health problems.

Under the Freedom of Information Act, Leonard was able to obtain a copy of the technical report documenting his findings at GE in 1979 and testified before a House Sub-Committee on Human Resources, chaired by Christopher Shays, on June 26, 1997. In his introductory remarks, he stated:

During four days of ground fighting [in the Gulf War], at least 300 tons of depleted uranium munitions were fired. An Army report describing research on hard target testing states that up to 70 percent of a depleted uranium penetrator can become aerosolized when it strikes a tank. Even if only 2 percent of the uranium burned up, then at least 6 tons of depleted uranium aerosol particles were generated – a huge amount – much of which would have become airborne over the battlefields. This amount in 4 days is more than 10,000 times greater than the maximum airborne emissions of depleted uranium allowed in the air over Albany in one month…

Unprotected U.S. service personnel inhaled and ingested quantities of depleted uranium particles into their lungs and bodies. They were never told about the health dangers of uranium particles and were given no means to protect themselves… After the war, many thousands of service personnel entered Iraqi tanks and armored vehicles that had been destroyed by depleted uranium penetrators, looking for souvenirs. They became contaminated… Twenty-seven soldiers of the 144th Army National Guard and Supply Company worked on and in 29 U.S. combat vehicles that had been hit by "friendly fire" and became contaminated with depleted uranium. They worked for 3 weeks without any protective gear before being informed that the vehicles were contaminated… U.S. troops were exposed to depleted uranium during [an ammunition storage area] fire and subsequent cleanup operations. They wore no protective clothing or masks during or after the fire. Approximately 3500 soldiers were based [there]…

This massive exposure to depleted uranium aerosol particles on the battlefield raises many questions about depleted uranium and how it might have caused at least some of the health problems now being experienced by Gulf War veterans…

It has been reported in *The Nation* that the Department of Veterans Affairs conducted a statewide survey of 251 Gulf War veterans' families in

Mississippi. Of their children conceived and born since the war, an astonishing 67 percent have illnesses rated severe or have missing eyes, missing ears, blood infections, respiratory problems, and fused fingers. The causes of these birth defects should be investigated. The human cost of using depleted uranium munitions in conflicts is not worth any short-term advantage if it permanently contaminates the environment and results in irreparable damage to our service personnel and causes genetic defects in their offspring.

Speaking as a World War II veteran, I am troubled about the health of Gulf War veterans and the seeming lack of concern shown by the Department of Veterans Affairs and the Army. They have refused to investigate the role of depleted uranium as a possible cause of Gulf War syndrome.

On a separate occasion, in November 1997, Leonard explained to me the problem of depleted uranium and its probable connection to Gulf War illnesses:

Thirty-six thousand Gulf War veterans are suffering from illnesses brought about during their service there, and nobody knows for sure what all the causes are. There's a possible connection with depleted uranium because depleted uranium metal was used in armor-piercing artillery. But the military refuses to acknowledge the dangers of the depleted uranium used in the Gulf.

The symptoms that the Gulf War veterans are suffering include chronic fatigue, rashes that come and go for no apparent reason, joint and muscle pain, headaches, memory loss, depression, lack of concentration, stomach problems, and many other maladies. Gulf War syndrome probably results from a combination of exposures, including depleted uranium.

The Committee on Government Reform and Oversight issued a report on the 7th of this month that was entered into the Congressional Record.[23] It's a devastating document, extremely critical of the Department of Defense, and contains the testimony of Gulf War veterans. And it has many more details that I've not seen elsewhere. It paints a totally different picture of what went on in the Gulf War. I genuinely think that the committee did everything it possibly could to bring the truth out on this issue in terms of the illnesses that Gulf War veterans have suffered, and it is a big step forward. It has confronted the highest levels of government for the very first time on this whole issue.

I was surprised to find out that the use of uranium metal in penetrators originated with the Nazis, in World War II. They were trying to develop an atomic weapon before us; they had access to uranium from the Belgian Congo, and they were trying to develop a centrifuge method for separating the isotopes. But they soon realized they were never going to beat us, because they had to keep moving their research sites around. They were being bombed out continually by the Allies. So they thought they might as well use the uranium in cannon projectiles and shoot them at our tanks...

People aren't going to drop dead the day after they're exposed to this. And the military realized that there was no way they could protect the soldiers on the battlefield; even today they refuse to admit that these particles can travel 26 miles. Fortunately I wrote that internal technical memorandum back in January 1980, which documents the actual numbers we measured.

I think the military will continue to deny the problem all along. I don't think they're going to change their view. They said so right before Shays' committee. A colonel from the Radio-Biological Research Institute got up and said, "I want to go on record that it is our position that depleted uranium is not dangerous." And he sat down.

To some degree, this kind of deception has been present in all wars, because one of the things the military does not want to do is to make public what's going on in a war while it's actually happening. A lot of information got to the American public during the Vietnam War which the military did not want publicized. But the Gulf War was totally different from all preceding wars in that there was absolute and total censorship of the press, and no reporters were allowed to accompany troops into the field. First time in our history. As a result, the public has been fed a lot of serious misinformation, and it's had an enormous effect.

The Gulf War is not an issue to most people anymore. It's over with. They killed any curiosity early in the game and did not let people in on what was really happening. Otherwise, it would have been a different picture entirely.

I can't imagine a military commander ducking responsibility like the Gulf War commanders did. On an NBC program three years ago, General Calvin Waller, who reported to Schwarzkopf, claimed that neither he nor Schwarzkopf knew anything about the health dangers of depleted uranium. Nobody told them. I find that hard to believe. Waller was being challenged by an NBC reporter with a document from a book called *Uranium Battlefields Home and Abroad,* which said, in effect, "Don't go into tanks that have been hit by depleted uranium munitions. They're radioactive." It gave instructions to troops on how to handle radioactive material, and those instructions came from the Army Chemical Command. And they were deliberately withheld until after the war.

Anybody who goes near Gulf War battlefields from now on risks contamination. Slowly, it'll become integrated in the soil. The uranium particles have a propensity for becoming attached to mineral particles in the soil. But they can become resuspended. Look at the helicopters flying over the battlefield, resuspending all that dust... The military knew that there was no possible way to prevent or clean up contamination on the battlefield, and therefore they decided to do nothing about it. And now it's been reported in Iraq that civilians there are showing the same symptoms as Gulf War veterans.

There are many, many first-rate scientists in the government laboratories. But the research is always directed toward a particular objective. I've tried to imagine how it came about that we actually used this deadly stuff in the first place. And I suspect the answer lies in taking the safety issue, for example, and breaking it down into little bits. You look for information that will reinforce your idea, so that you can go ahead and use it. You give it to a scientist, and you limit what he can tell you. It's like a lawyer, exactly analogous to a lawyer in a courtroom, trying to limit what a witness can say regarding a particular case. You're not free to say anything you want to. You might even suspect what the big picture is, and you probably do, because if you didn't you aren't very bright. But you're only asked one very specific thing, and you have to respond to that question only, and that's it. Everything else is extraneous.

I think one of the great fears in the scientific community is that you're going to look foolish if you make a statement and you're later proven to be wrong. I agonized over going public a long time before deciding that I *had* to go public. This was before the war began, based on what I knew about the transport of uranium particles. I was an insider. Yet I wrote this unclassified report, which I was glad to see released under the Freedom of Information Act. It's a smoking gun. I knew that it was an important discovery at the time. What I didn't know exactly was how it fitted in...

People want to forget about the war. Every war is like that. I think it's very important for anybody who's contemplating a military career to look at what's happened to the veterans of past wars. I've been asked, "Which was worse, to drop the atomic bombs or to drop the firebombs?" And I've come to the conclusion that it's a meaningless question. The real question is: What about war itself? I've come to believe, as a person and as a scientist, that war is immoral and cannot be tolerated. We have to eliminate war. If we don't do it, the human race will not survive. That's my conclusion.

9 God on Our Side

Here we were

not only winning this war, but we were routing the enemy – absolutely routing the enemy – and yet our casualties were practically…nonexistent. You know, that kind of made you feel that God was on your side…

General Schwarzkopf, Supreme Commander, Armed Forces, Gulf War

From the Crusades to the Gulf War, leaders of both church and state have used God's name to justify the most awful butcheries in world history. There is no more effective catalyst, no finer inspiration in battle, than an appeal to the very Maker of man. When a nation and its armies stand convinced that they are aligned with the divine cause, wavering is banished, and the bloodiest carnage can proceed unhindered.

The invocation of the *jihad,* the holy war, is a ritual that has been refined through the ages. The Bible's Old Testament is rife with wholesale slaughter in the name of a just and righteous God, carried out at the hands of a people with God on their side. Yet buried deep within the clamor is the disconcerting whisper of God, speaking to King David of all that has gone before: "You have killed too many men in great wars. You have reddened the ground before me with blood: so you are not to build my Temple."[1]

Down to the present day, leaders of church and state alike have ignored this whisper. It is as an aberration, a blip in the stream of consciousness that persistently aligns the Creator with man's destruction of creation. And the notion of killing for God has taken forms that are ever more subtle, more presentable, and perhaps more effective.

Like his predecessor, Ronald Reagan, George Bush understood well the tremendous weight that "God-inspired" intervention carried with the American people. In his inaugural address in January 1989, the ex-CIA head glowed piously:

My first act as President is a prayer. I ask you to bow your heads. Heavenly Father…write on our hearts these words:"Use power to help people." For we are given power not to advance our own purposes, nor to make a great show in the world, nor a name. There is but one just use of power, and it is to serve people. Help us to remember it, Lord. Amen.

That such sentiments swayed the hearts of the religious establishment is certain; that Bush used his power to "help people" is not. I will never forget seeing the 1992 Academy Award-winning documentary, The Panama Deception, in a nineteenth-century movie house in the small town of Redhook, New York. I watched, unbelieving, as smuggled footage of the actual U.S. invasion of Panama was dubbed over glowing clips of U.S. government officials praising the "surgical" nature of Bush's crusade to capture Noriega.

Whole residential areas of Panama City were aflame. In a brutally enforced media blackout, a journalist was shot by U.S. forces when he stepped over the tape around the "no see" zone. U.S. Marines dragged screaming Panamanian citizens out of their houses, laid them face down in the street and shot them in cold blood. And in the aftermath of the invasion, which the U.S. government official claimed had taken the lives of only a very few civilians, Panamanian workers were shown uncovering mass graves of executed citizens. A number of corpses had holes burned through their bodies – caused, according to the testimony of eyewitnesses, by experimental laser weaponry.

It was impossible to watch such a documentary and not arrive at the conclusion that the highly publicized hunt for

Noriega was nothing more than a smoke screen, a cover-up of a blow to the Panamanian Defense Forces protecting the canal.*

Apparently troubled by his somewhat messy "victory" in Panama, Bush proceeded to lead America into yet another massacre – this time of hundreds of thousands of Iraqis in the Gulf War. "This will not be another Vietnam," he assured the American people shortly before the attack. "Our troops will have the best possible support in the entire world. They will not be asked to fight with one hand tied behind their back." In other words, the U.S. military would be free to unleash its full firepower against a vastly inferior nation without political constraint; they would be free to "let the Iraqis have it." They did.

In October 1992, George Bush sermonized once again on the divine sanction of American endeavors: "We believe in...the Judeo-Christian heritage that informs our culture... [We must] advocate the recitation of the Pledge of Allegiance as a reminder of the principles that sustain us as one nation under God."

As shocking as these words may seem, their spirit is not new. Every war in this century has been defended in no less scandalous a fashion, and with similar appeals to duty and moral conviction. Each time, thousands of young men have responded, many recognizing too late the hypocrisy that would be inherent in a God who blesses war.

* The Panamanian Defense Forces was one of the primary targets of the invasion. The Canal Treaty signed by President Carter and General Torrijos states that in order for the canal to return to Panama in 1999, Panama must have the means to protect it. Following the invasion, Panama amended its constitution to abolish the P.D.F.

THE CONTROVERSIAL *Enola Gay* exhibit in the Smithsonian's Air and Space Museum, in Washington, DC, was originally intended to be a comprehensive exploration of the dropping of the atomic bombs on Hiroshima and Nagasaki. But the first draft of the guide was so honest that it was pulled by the Senate Rules and Administration Committee in response to the outcry of organizations like the Veterans of Foreign Wars and the American Legion, who wanted "anti-American" material expunged.

Senator Ted Stevens of Alaska, who chaired the committee, framed the directive to the Smithsonian most succinctly: "I don't think you have any authority to display an exhibit questioning U.S. use of the atomic bomb..."[2]

The committee got its way. What I saw at the *Enola Gay* exhibit was history sanitized beyond belief. The first thing that confronts the eye is the huge vertical stabilizer and rudder of the restored B-29, gleaming silver, adorned with a large black "R" inside a perfect circle. After reading an exhaustive history of the development of the B-29 and the construction of the *Enola Gay,* visitors follow a corridor through displays of propellers, engine cowlings, and piping.

The focal point of the display is a large space containing the entire forward fuselage of the aircraft. It looks so pristine; the shining silver of the riveted plating and the polished black of the interior controls give no hint of atrocity, no evidence of the monstrous devastation once unleashed upon mankind through the yawning bomb-bay doors. As the visitor walks around the nose of the bomber, a glass casing beneath its belly comes into view. Inside is a replica of "Little Boy" itself: a nine-foot cylinder with square fins, painted a flat army green.

The superficiality of the exhibit is at once sinister and ridiculous: museum-goers are reassured that the restoration materials are "authentic," and that the replica of the bomb itself poses no risk of radiation. A small plaque states only that the bomb casing inside the glass is a dummy used for "training."

The tour ends in a small movie theater, where members of the American bombing crews are interviewed on film before and after the destruction of Hiroshima. The reel is short and to the point: over footage of preparations on the morning of August 6, 1945, we hear a chaplain calling on God to protect the men of *Enola Gay* as they "make the world safe for peace and for democracy – in the name of Jesus Christ, Amen." One member of the *Enola Gay* crew reflects, almost unbelievably, "We succeeded in bringing the carnage to an end."

The message the exhibit delivers is unmistakable: Americans can take pride in their nation's role of global "redeemer." But I couldn't help thinking of Father George Zabelka, a Catholic chaplain with the U.S. Army Air Force, who was stationed on Tinian Island in the South Pacific in August 1945. Zabelka served as a priest for the airmen who dropped the atomic bombs on Hiroshima and Nagasaki, and his long and painful inward journey began when he counseled an airman who had flown a low-level reconnaissance flight over the city of Nagasaki shortly after the detonation of "Fat Man."

Not much could be seen from bombing altitude, but a low run directly above the city revealed a scene out of Dante's

Inferno. Thousands of scorched, twisted bodies writhed on the ground in the final throes of death, while those still on their feet wandered aimlessly in shock – flesh seared, melted, and falling off – like the walking dead. The crewman's description raised a stifled cry from the depths of Zabelka's soul: "My God, what have we done?" And over the next twenty years, he gradually came to believe that he had been terribly wrong, that he had denied the very foundations of his faith by lending moral and religious support to the men who had wreaked such hellish destruction.

When asked if he had known that civilians were being destroyed by the thousands in air raids, long before the atomic bombing, he replied:

Oh, indeed I did know, and I knew with a clarity that few others could have had…

The destruction of civilians in war was always forbidden by the Church, and if a soldier came to me and asked if he could put a bullet through a child's head, I would have told him, absolutely not. That would be mortally sinful. But in 1945 Tinian Island was the largest airfield in the world. Three planes a minute could take off from it around the clock. Many of these planes went to Japan with the express purpose of killing not one child or one civilian but of slaughtering hundreds and thousands and tens of thousands of children and civilians – and I said nothing…

I never preached a single sermon against killing civilians to the men who were doing it… I was brainwashed! It never entered my mind to publicly protest the consequences of these massive air raids. I was told it was necessary – told openly by the military and told implicitly by my church's leadership. To the best of my knowledge no American cardinals or bishops were opposing these mass air raids. Silence in such matters, especially by a public body like the American bishops, is a stamp of approval.[3]

How did Zabelka come to change his views so radically? At a 1985 Pax Christi conference, he told his story in words filled with urgency and power:

I worked with Martin Luther King, Jr. during the Civil Rights struggle in Flint, Michigan. His example and his words of nonviolent action, choosing love instead of hate, truth instead of lies, and nonviolence instead of violence stirred me deeply. This brought me face to face with pacifism… active nonviolent resistance to evil. I recall his words after he was jailed in Montgomery, and this blew my mind. He said, "Blood may flow in the streets of Montgomery before we gain our freedom, but it must be our blood that flows, and not that of the white man. We must not harm a single hair on the head of our white brothers…" I struggled. I argued. But yes, there it was in the Sermon on the Mount, very clear: "Love your enemies. Return good for evil." I went through a crisis of faith – a crisis of faith, not a crisis of priesthood; that's simple! A crisis of faith. Either accept what Christ said, as unpassable and silly as it may seem, or deny him completely…

For the last 1700 years the church has not only been making war respectable: it has been inducing people to believe it is an honorable profession, an honorable Christian profession. This is not true. We have been brainwashed. This is a lie…

War is now, always has been, and always will be bad, bad news. I was there. I saw real war. Those who have seen real war will bear me out. I assure you, it is not of Christ. It is not Christ's way. There is no way to conduct real war in conformity with the teachings of Jesus. There is no way to train people for real war in conformity with the teachings of Jesus…

So the world is watching today. Ethical hairsplitting over the morality of various types of instruments and structures of mass slaughter is not what the world needs from the church, although it is what the world has come to expect from the followers of Christ. What the world needs is a grouping of Christians that will stand up and pay up with Jesus Christ. What the world needs is Christians who, in language that the simplest soul could understand, will proclaim: the follower of Christ cannot participate in mass slaughter. He or she must love as Christ loved, live as Christ lived and, if necessary, die as Christ died, loving one's enemies… When Christ disarmed Peter he disarmed all Christians…[4]

Zabelka died in 1992, without many Americans ever hearing his extraordinary story. But in reflections confided to at least a few listening ears, he unmasked what he saw as the moral and religious hypocrisy of the atomic bombings with unparalleled conviction. He spent his last years elaborating on the absolute incompatibility between the teachings of war and the teachings of Christ.

I painted a machine gun in the loving hands of the nonviolent Jesus, and then handed this perverse picture to the world as truth. I sang "praise the Lord" and passed the ammunition…All I can say today is that I was wrong. Christ would not be the instrument to unleash such horror on his people… I say with my whole heart and soul I am sorry.[5]

Maureen Burn, whose husband Matthew fought in World War I, remembers well how the British newspaper editors and clergy alike exhorted the masses to join in the "war to end war."

Matthew was an Anglican choirboy in 1914, but he ran away from school and enlisted. His parents went to fetch him because

he was not yet of age, but were dissuaded by an officer who told them, "With your boy, this war is a holy crusade. He will just run away again."

Soon Matthew was in France, where he experienced the full horrors of trench warfare; he came home a changed and bitter man. Recently, Maureen told me:

Matthew declared he was not going to do a thing for a rotten society, where even the clergy preached war to their young people. He said he would just be a tramp and wander from one poorhouse to another when it was too cold to sleep outdoors.

But his older sister, who was a dedicated nurse, told him, "You are a victim of this rotten society, but what good could you do for other victims of this rotten society if you were just a tramp? It would be quite a different thing if you would study to be a doctor." That convinced him. He went to Edinburgh University, and after finishing the basic course went on to study tropical medicine.

I remember a walk with Matthew that changed my outlook on life. He asked me about my father, and when I told him Father had been a clergyman and a missionary, he said, "The clergy are just parasites. They should do a hard day's work." I was shocked. I asked him why he said this, and he said that the clergy had conscripted all his friends and they had all been killed. Then he said the only time he had ever seen the Christian spirit was in the frontline trenches, when the crudest Tommy would give his last crust to a dying companion. "But," he added, "that spirit never lasted behind the lines, when the fellows got into safety."

I left the church when I realized that it went along with war. I wanted to join the Quakers, the only historical peace church in Britain, but Matthew said, "The Quakers are just the Red Cross behind the capitalist firing lines!" He explained that the Quakers were the first bankers, and their banks were heavily involved in big business and war. So I didn't join them, although I attended their meetings.

I threw myself into the peace movement, but Matthew, who was more of a realist than I was, was quite skeptical about it. He had always wanted to go to Germany and meet Germans who had been in the front line, but he knew no German. I had studied German in school, so we took a walking tour of the Black Forest, putting up in the little village inns where the local people came in for a drink of wine in the evenings. He could spot an ex-frontline man, and he would ask me to tell this or that man that he recognized him as such. Then he would offer him a drink. At first the German would be a bit taken aback, but then he would become friendly when he realized that Matthew had been through the same experience and hated war. They always ended up saying, *"Krieg ist Wahnsinn"* (war is madness).

On Armistice Day he always disappeared. I don't know where he went. He thought it an insult to the dead to have a big military parade at the cenotaph where the Unknown Soldier was buried…

EBERHARD ARNOLD, a German pastor and theologian, also blessed his country's war efforts. A convinced young Christian from an intellectual, upper middle-class home, he received a doctoral degree in philosophy from Erlangen in 1909, and soon he defied his family's social and academic standing by immersing himself in social issues, such as the plight of the poor. In 1914, when the war first broke out, he had reported eagerly for duty.

Though soon discharged for health reasons, Arnold felt that the war had a divine import: it was a call to a deeper Christianity. In 1915 he joined the Relief Action Committee for Prisoners of War, which provided encouraging literary material to pris-

oners of war and to wounded soldiers in military hospitals. In an article he wrote:

We are living in a tremendous time, and we cannot be thankful enough for what God has brought about by enthusing men for the fatherland and by lifting men's hearts up to God. A love has arisen in our nation that does not move in sweetly sentimental ways, but rather gives itself essentially in strength. It is a love shown in action, ready to sacrifice itself where so much has been given, a love that lays down its life on the altar of the fatherland, for the emperor, for relatives, and compatriots...

In this war, too, everything depends on love remaining in us, for it is love that impels us to give our lives for our brothers. The soldiers who know that Jesus died for them will show the greatest endurance. They will most confidently and serenely risk their lives for their brothers. There is no holier incentive for a soldier than this fact: Jesus died for me, therefore I owe it to give my life, too. Our nation will be victorious if the roots of our strength lie not in hatred but in love, if we have the impulse of love for our enemies.

At present the soldiers have to overcome the lies with fire and the sword, but the driving power of our military action has to be the well-tried German love; more than that, it has to be the love of Jesus, which wants to include the love for England, for the peoples under England's yoke, and for all nations groaning for love...[6]

As the war dragged on, however, Arnold's convictions were increasingly challenged by the suffering he saw around him:

If anyone had an eye for the suffering of the people in the city, how grieved his soul would be, seeing the terrible suffering caused by misled longing and exploited nostalgia! It is deplorable how many souls are stripped of their innocence, as when the bloom is brushed from a butterfly's wings. Oh, what sorrow grips us at the sight of such deluded yearning.

But we can offer a hope to all men, even to those most deeply wounded, to every disappointed soul; the hope that all who have been disappointed by false love can find true love.[7]

By 1917, his tone had changed completely. No longer did he speak of a "well-tried German love," but rather of the evil that had wrought such protracted, senseless carnage:

These years we are living in have shown more and more what power hatred has among men and in what a frightening way the love of men is destroyed by wickedness and lawlessness; how men's fury, goaded on by the lawless acts of some other faction, drives them to fanatical extremes of hatred. I will not recall how the nations work themselves up to strangle each other, and the classes within nations in turn work each other up to hate and murder one another; how lawlessness among the social classes keeps fanning the hatred to a blazing fire so that no one can tell how horrible the end will be. Yes, war reveals the evil of hatred and murder and lawlessness – the evil of love grown cold.[8]

The war finally over, Arnold went through a period of deep soul-searching to find what God really demanded of him as a Christian. His previous belief in a "personal" Christianity, focused on saving souls amidst the upheaval of world events, seemed empty, inadequate. He wrote:

We saw the condition of the men who came home. One young officer came back with both his legs shot off. He came back to his fiancée, hoping to receive the loving care he needed so badly from her, and she informed him that she had become engaged to a man who had a healthy body.

Then the time of hunger came to Berlin. People ate turnips morning, noon, and evening...I saw a horse fall in the street; the driver was knocked aside by the starving people who rushed in to cut pieces of the meat from the still warm and steaming body so that they would have something to take home to their wives and children...

After such experiences and after the enormous revolutionary changes, which resulted in turning elegant apartments with parquet floors and huge reception rooms over to working-class families, I realized that the whole situation was unbearable. I was told...that a high state official demanded of me that I should be silent about the social problems, about the war, and about the suffering that cried out to heaven...[9]

By now convinced that war, in any form and for any reason, was utterly wrong, Arnold felt led to abandon not only war, but the whole social and economic order that brought it about:

Love does not take the slightest part in hostilities, strife, or war, nor can it ever return curses and hate, hurt and enmity...Love is not influenced by any hostile power. No change of circumstance can change the attitude of Jesus and his followers; he does nothing but love, make peace, wish and ask for good, and work good deeds. Where the peace of Jesus dwells, war dies out, weapons are melted down, and hostility vanishes. Love has become boundless in Jesus; it has achieved absolute sovereignty.[10]

Like Eberhard Arnold, Adolf Braun was called up for duty when the war broke out in 1914. The son of a Baptist minister, he had entered the military service in 1913 and was a member of the Prussian Garde Artillery Regiment in Spandau, Berlin. He was one of thousands of German soldiers gripped by a sense of "sacred duty to defend the fatherland," and he and his comrades left their hometowns amidst enthusiastic singing, the strewing of flowers, flag-waving by families and friends.

Churchmen prayed for victory and for the safe return of soldiers. Adolf was ebullient: "With God, for king and fatherland!"

He fought on both fronts, in France and Russia, where he became known for his valor in jumping out of trenches to throw hand grenades at the enemy, which resulted in his being wounded several times. Then in July 1915 he was injured by shrapnel that entered his head behind the ear.

He was in critical condition for weeks and remained hospitalized for five months. But in February 1917 the Army was desperate for soldiers – even wounded ones – and Adolf was drafted a second time. He was sent again to France, for a final desperate effort in a war that was by then clearly lost.

It was in this last wrenching year that Adolf was profoundly affected by the insanity of it all. Once, when stationed with his men in the woods, he looked down through the fog and smoke to see a French observer in another tree, not far away. Bullets whistled around him in the cold rain, and he thought of home, where his bride and loved ones were surely praying for him. Then he looked over at the Frenchman and wondered, "Are his loved ones praying for him, too?" He was shaken by this thought and decided that if he ever got out alive, his life would be very different. The losses during this battle were particularly heavy, and when he came down from his post in the tree, he found that of his entire company he was the only one left alive.

One night his men were ordered to attack a company of French soldiers camped in a small grove of trees. It was to be a surprise attack, and in order not to wake the French, no shot was to be fired: they would engage only in hand-to-hand bayonet fighting. When it was over, the men were so exhausted that they fell asleep right next to the cannons. But Adolf and a comrade were awakened by moaning, and after following the sound they found a young French officer crying in pain. Adolf bent over to console him, but the Frenchman pointed weakly to his pack. In it was a bloodstained Bible and a photograph of his family.

As Adolf read to the dying man from the Bible, he was suddenly overcome by the madness of the whole situation: here he was, trying to comfort a man he had ordered his men to kill! Before dying, the enemy soldier gave Adolf his Bible and begged him to notify his wife of his death.

When Adolf returned home, he was unable to settle down into the comfortable existence of a "normal" life. He was offered promotion to officer's rank, an honor to be bestowed by the Kaiser himself, but he refused; had it not been for his good connections, that snubbing could have cost him his life. The war left Adolf shattered. He was determined to see that there would never be war again. His war bride, Martha, felt as he did, and together they sought a way of life that would be completely different from the one that had led them and their countrymen to war.

Adolf and Martha began to reach out, helping to alleviate the tremendous need they saw all around them. Adolf began visiting men in prison, assisting their families with packages of food and clothing. He spoke out against the war with

a candor grounded in his own experiences and argued that talk of a "hero's death for the fatherland" was meaningless and hypocritical. All this earned him more than a few enemies, and his friends worried for his safety and even his life.

Eventually, his search for peace and social justice led him to a Christian community where the needs of each member were cared for by all. Years later, after being driven from Germany by the Nazis and forced to emigrate to Paraguay, Adolf wrote to his family back home:

You will perhaps recall how in 1914, as a young Christian patriot, I went into battle "with God for king and fatherland." But like hundreds of thousands of my compatriots, I had no inkling of the powers that dictate world events and manipulate things for their own interests; how the nations are played off, one against the other; how poets – surely in poetic blindness – make brutality appealing; how the representatives of almost every religion bless weapons with their false intercessions and their propaganda for a "Christian" state – something that cannot be; how on both sides soldiers are consecrated for a sacrifice that is completely misunderstood. And finally, how the holy sacrificial death of Jesus is compared – even equated – with the sacrifice of soldiers, although in reality none of them is ready to sacrifice himself, but only the life of the enemy…

Your letter was the first sign of life after the dreadful time that should now lie behind us, if men only knew how to make true peace! But that is of course impossible, as long as we do not courageously face up to the real motives that lead to war; impossible, when even in times of so-called peace everyone is battling everyone else; impossible, as long as stubbornness, selfishness, and private property are protected and supported by Christian morality…[11]

Soldiers like Adolf were not the only ones to indict the church for its support of the Great War – in fact, some of its sharpest condemners were found among the clergy themselves. In this regard, few public addresses surpass that of Harry Emerson Fosdick, an American military chaplain who on November 12, 1933, delivered a sermon entitled "The Unknown Soldier," a speech later printed in the U.S. Congressional Record:

You may not say that I, being a Christian minister, did not know [the Unknown Soldier]. I knew him well. From the north of Scotland, where they planted the sea with mines, to the trenches of France, I lived with him and his fellows – British, Australian, New Zealander, French, American. The places where he fought, from Ypres through the Somme battlefield to the southern trenches, I saw while he still was there. I lived with him in dugouts, in the trenches, and on destroyers searching for submarines off the shores of France. Short of actual battle, from training camp to hospital, from fleet to no-man's-land, I, a Christian minister, saw the war. Moreover, I, a Christian minister, participated in it. I, too, was persuaded that it was a war to end war. I, too, was a gullible fool and thought that modern war could somehow make the world safe for democracy. They sent men like me to explain to the Army the high meanings of war and, by every argument we could command, to strengthen their morale. I wonder if I ever spoke to the Unknown Soldier.

One night, in a ruined barn behind the lines, I spoke at sunset to a company of hand-grenaders who were going out that night to raid the German trenches. They told me that on the average no more than half a company came back from such a raid, and I, a minister of Christ, tried to nerve them for their suicidal and murderous endeavor. I wonder if the Unknown Soldier was in that barn that night…

You here this morning may listen to the rest of this sermon or not – as you please. It makes much less difference to me than usual what you do or think. I have an account to settle in this pulpit today between my soul and the Unknown Soldier.

He is not so utterly unknown as we sometimes think. Of one thing we can be certain: he was sound of mind and body. We made sure of that. All primitive gods who demanded bloody sacrifices on their altars insisted that the animals should be of the best, without mar or hurt. Turn to the Old Testament and you find it written there: "Whether male or female, he shall offer it without blemish before Jehovah." The god of war still maintains the old demand. These men to be sacrificed upon his altars were sound and strong. Once there might have been guessing about that. Not now. Now we have medical science, which tests the prospective soldier's body. Now we have psychiatry, which tests his mind. We used them both to make sure that these sacrifices for the god of war were without blemish. Of all insane and suicidal procedures, can you imagine anything madder than this, that all the nations should pick out their best, use their scientific skill to make certain that they are the best, and then in one mighty holocaust offer 10,000,000 of them on the battlefields of one war?

I have an account to settle between my soul and the Unknown Soldier. I deceived him. I deceived myself first, unwittingly, and then I deceived him, assuring him that good consequence could come out of that. As a matter of hard-headed, biological fact, what good can come out of that? Mad civilization, you cannot sacrifice on bloody altars the best of your breed and expect anything to compensate for that...

When you stand in Arlington before the tomb of the Unknown Soldier on some occasion, let us say, when the panoply of military glory decks it with music and color, are you thrilled? I am not – not anymore. I see there the memorial of one of the saddest things in American history – from the continued repetition of which may God deliver us! – the conscripted boy...

If I blame anybody about this matter, it is men like myself, who ought to have known better...

The glory of war comes from poets, preachers, orators, the writers of martial music, statesmen preparing flowery proclamations for the people, who dress up war for other men to fight. They do not go to the trenches. They do not go over the top again and again and again...

I will myself do the best I can to settle my account with the Unknown Soldier. I renounce war. I renounce war because of what it does to our own men. I have watched them coming gassed from the frontline trenches. I have seen the long, long hospital trains filled with their mutilated bodies. I have heard the cries of the crazed and the prayers of those who wanted to die and could not, and I remember the maimed and ruined men for whom the war is not yet over. I renounce war because of what it compels us to do to our enemies, bombing their mothers in villages, starving their children by blockades, laughing over our coffee cups about every damnable thing we have been able to do to them. I renounce war for its consequences, for the lies it lives on and propagates, for the undying hatreds it arouses, for the dictatorships it puts in the place of democracy, for the starvation that stalks after it. I renounce war, and never again, directly or indirectly, will I sanction or support another...[12]

THE STAND OF AMERICAN churches in regard to the Vietnam War was hauntingly familiar. As in the past, there were churches that protested the war, but for the most part, institutional religion once again repeated its practiced dance with governmental might, solemnly acceding to carnage in the name of democracy. Jerry Voll, a minister in Lancaster, Pennsylvania, during the late 1960s, remembers:

The Franklin and Marshall campus was a real seedbed for the protest movement. After a Sunday morning service, a young man from the college came up to me – he was probably around twenty-one – and said, "Reverend, does your church believe in war, or does your church take a stand against war?" I looked at him and I was speechless, but I knew what my answer had to be, and I didn't like it. I said, "No, the church does not take a stand against the war – you can believe whatever you want to believe." At that very moment, I knew this was wrong.

That young man convinced me to do something about the war, and because I was on the local council of churches responsible for community action, I recommended at the next meeting that we support and recognize the young people who were doing such a good job making us aware of these issues. One minister after another said they couldn't, for one reason or another, and in the end I just got up and left. That was about the end of my career as a minister.

Author Gloria Emerson met more than just apathy in her interviews with Vietnam-era military chaplains. One, a certain Colonel Beaver, said, "[Jesus] was wrong. Greater love hath no man than when he lay down his life for a *stranger*... That is what the United States is doing in Vietnam." Another, Major Emlyn Jones of the Church of the Brethren, recalled his tour in Vietnam with vehemence: "It gave me sorrow," he said, "but most of all it gave me a tremendous hatred of communists. Man! I hated those spastics!"[13]

ON A COLD NOVEMBER day in 1997, I drove to a farm in Marlboro, New York, hoping for an interview with renowned peace activist Tom Cornell. Garbed in a faded sweatshirt and barnyard

boots, Tom greeted me with characteristic warmth and humor. After the requisite tour of the rugged and somewhat swampy farmland, we headed for the Cornells' kitchen, where we pored over pictures and news clippings of bygone arrests and demonstrations and talked peace and politics over ziti, coffee, and homemade bread.

When we finally retired to the living room by early afternoon, it was nearly dark outside, threatening snow. Tom looked out into the winter sky, and began:

I was much too young for World War II – I was eight years old when Congress declared war in 1941 – but that war was quite an experience, even for a kid, because everybody loved it in a way. It really bound the country together. We were just getting out of the Depression. People of my class and kind thought that Franklin Roosevelt was the fourth person in the Trinity. Irish-Italian Democrats, we all thought the world of FDR. We went to war with great enthusiasm, like most people. Maybe with more enthusiasm than many. But I saw how the older people worried. One of my uncles became a temporary alcoholic, just from worrying that his oldest boy, who was in the tank corps with Patton in North Africa, was going to get hit.

Well, our boys came home. I was still too young by a hair for the war in Korea, but for some reason the war in Korea never excited anybody that I knew. I was more excited about the Trojan War. I was learning Greek and reading the Iliad and the Odyssey. I was doing Virgil in Latin. So the windswept hills of Troy meant far more to me than anything in Asia.

I registered for the draft when I was eighteen. Everybody did; you just did it. So I registered and thought nothing about it. One of my buddies had applied for the Marine Corps officers training program, and they were offering to pay his way through college. Now it looked like I was going to have a hard time paying for college, and I thought, I'll give it a try. So I presented myself to the Marine Corps and they said, "You're a little too scrawny. Put on enough weight and we'll take you." I just ate and ate for a month, but I couldn't put on quite enough weight. They said that they'd take me as a recruit, but not as an officer. I told them I was born officer class – that would never do, I was just brought up that way. So I gave up that idea.

I always just assumed that a Catholic couldn't be a conscientious objector to war. At the pre-induction physical examination they ask, "Are you a conscientious objector?" I said, "No." There was a kid behind me, though, who couldn't speak English, only Italian, so they asked me to translate for him. I couldn't figure out how you say "conscientious objector" in Italian. Now I know how to say it, but then I didn't. So I said to him simply, "Are you Catholic?" And he said, "Yes, I'm Catholic." And I said, "No, he's not a conscientious objector" – because in my mind the two were mutually exclusive.

Years and years later I remembered an episode from the third grade. This was right in the middle of World War II. Miss Coughlin – I had a crush on her – brought in the front page of the Bridgeport Post, and she showed it to the class. There was a photograph of a young Jehovah's Witness who had just gone to Danbury Prison. She said he'd gone to prison because he refused to kill, and that he was a conscientious objector to war because of his religious beliefs. Miss Coughlin was Catholic and I was Catholic, and I remember saying to myself, "Gee, I'm glad I'm Catholic and will never have to face a question like that." Years later I realized that yes, I did.

I spent quite a bit of time trying to figure out if I really was a conscientious objector. I knew that very few Catholics had taken this stance. I went for help to a theologian. He said, "You know, I have to agree with you that this is a legitimate Catholic position, but I don't think I could take it myself." I said, "Why, Father?" And he said, "Because I couldn't have taken it in the Irish War," which was the war against England. To him that was the "just war." Everybody has his "just war."

I got my first draft card shortly after I applied, right out of college, in 1956. In high school I was given a student deferment, 1-S; in college I had a

2-S. When I got out of college I lost the deferment and got a 1-A. I rejected the 1-A and told them I would not be available. (1-A means you're available for service as a combatant.) I insisted upon having a hearing as a conscientious objector. They said, "Okay, we'll make a compromise with you. We'll give you a 1-AO, which means that you're available for induction into the military as a noncombatant. You don't have to carry or transport weapons, and you can be a Christian witness all you like. But you're in the Army." I said, "No, I can't accept that. I won't be under military authority at all. And besides, if I put that uniform on, the uniform says something; it says something I don't believe in." They said they couldn't buy that, but then nothing happened for four years. The fighting in Korea had ended, and Vietnam was almost a decade away.

In 1960, they gave me a 1-O. I'm twenty-six, and I said, "Wow, this is great. I don't have to worry about going to prison anymore, and I don't even have to do alternative service, because I'm too old to be drafted." I kept the draft card in my wallet to prove that I was old enough to buy a beer. That's the only thing it was ever useful for.

On one occasion that summer, my buddy Loren Miner and I decided we should go have a beer. So we went, and the bartender said to me, "Are you twenty-one years old?" And I said, "No, sir. I am twenty-six years old." "Can you prove it? Where's your draft card?" "Here's my draft card." And I handed it to him. He looked at it, he handed it back, and I said, "I don't want it." I realized that I really didn't want the card. So I had the beer, and then I said, "Loren, you know what I think I'm going to do? I'm not going to be part of this system anymore. I'm going to 'unregister.' And I'm going to burn this card right here and now." So I burned it in the ash tray on the bar at the Goldenrod on Bank Street, had another beer, and went home.

In 1965 the antiwar movement began to heat up. Young people went to Washington that spring, 1500 or so, to a very conservative demonstration that was sponsored by the Students for a Democratic Society. On August 28,

I think it was, *Life* magazine printed a full-page black-and-white photograph of a young man burning a draft card at the Whitehall Induction Center. This young man happened to be the great-great-grandson of a signer of the Declaration of Independence, and he was a good-looking white boy with a blond crew cut, looking intently into a burning draft card. They had a copy of the magazine in the Senate barbershop, and when the senators went to get their hair cut and saw the picture they just went berserk, or, as they say today, they went ballistic. They said, "Look at these kids doing these things! This can't be tolerated! This is America, after all!" So they rushed through a law saying that you'd get five years and a $10,000 fine if you burned your draft card.

It was clear what they intended to do. They intended to use the law to stifle dissent. I really love the democratic traditions of this country, and the one I love the most is the freedom to dissent. This is the American tradition.

Well, didn't we then have an obligation, those of us who had burned draft cards and suggested the idea to others, to defy this law? It was immediately clear to me that we did. I didn't have any draft cards left to burn, so I wrote to my draft board. I said, "Ladies and gentlemen of the draft board, I am not in possession of a registration or classification certificate. I burned them years ago. Would you please send me duplicates." And they did. And we had another ceremonial draft card burning...

I don't know what's going to happen in the future. We may never have a draft again, because war has become so sophisticated, technologically, that you no longer need millions of men. Now we have missiles that can turn a 90-degree angle in midair. If we have a prolonged war, even without a draft, the public is going to be much less apt to write a blank check. People aren't going to buy it. They know better. At least I hope that there are enough people asking questions now, so that it'll be harder to wage a war that lasts for any length of time. Just to start asking those questions is the most important thing for young people.

AMONG "GOD'S TROUBLEMAKERS," few of his contemporaries are better known than Philip Berrigan. A World War II veteran and former Josephite priest, he has been been fighting the military-industrial complex since the early 1960s.

On February 12, 1997, Ash Wednesday, he and fellow members of the activist Prince of Peace Plowshares boarded the USS The Sullivans, a Navy guided-missile destroyer, at the Bath Iron Works in Maine. Armed with hammers and bottles of blood, their goal was to penetrate the ship's control room and "disarm" the main console. One of the men quickly ran past the gathering security personnel and managed, somehow, to find the control room. For several minutes, he hammered away at the key pads and controls, finally pouring his blood over the cracked display panels. When he was finally apprehended, he knelt down on the deck in front of his captors to say a short prayer for them.

Meanwhile, Philip Berrigan had been stopped on the top-side deck. A frantic young security officer brandished a beretta at him, cursing and screaming. Philip, trying to calm him, said, "Son, put that thing away. That thing you've got in your hand is the source of all that's wrong with this country." The young man continued to threaten Philip. "You gonna shoot me, huh?" Philip asked softly. "That's what the Navy's all about. Don't you get dragged into it too." "Na, I wouldn't shoot ya," the young man retorted, glaring. "I'd come upside your head and knock you out."

I visited Philip at the Cumberland County Jail in Portland, Maine, eight months later. A warm, hearty man, he was soft-spoken and meek, dressed in bright orange prison pajamas and with a smile that defied that overheated and harshly lit cubicle where we met. Shortly afterward, he was sentenced to two years in prison, followed by two years of probation, and a fine of $4,703.89 for the damage incurred to the warship. The federal district judge made a point of saying that he was not passing judgment on the morality, propriety, or sincerity of Philip's actions, but that his views on nuclear weapons could not justify violation of the law.

His personal journey is remarkable. As a young man, he marched into battle in World War II with the patriotic fervor of a trained killer. As a member of a field artillery outfit, he served first in Normandy and later fired on German submarine emplacements at Lorient and Brittany, moving across Northern France into the Low Countries and then on toward Germany. He came home a "victory boy," restless but idle, basking in the fading light of glory. The fifties found him in the Deep South, a Josephite priest teaching black children and learning from them, firsthand, their oppression and struggle to survive. Such experiences fueled a growing discontent with the status quo and revealed to him the very deep connections between racism and war. By 1967, he joined his brother Daniel, a Jesuit priest, in burning draft files in Catonsville, Maryland – the first link in a ponderous chain of actions taken to arouse the American mind and to protest the manufacture of weapons of death. Since 1967, Philip has been arrested for civil disobedience more than one hundred times, and has spent a total of more than seven years in prison.

The Irish twinkle in his eye warms the heart but belies the deadly seriousness of his convictions. There is an air of craziness about him, too, but one calculated, it seems, to counteract the craziness of a society gone mad.

I was a very good killer. I was a superior marksman – I knew how to handle a bayonet, an automatic rifle, a BAR (Browning automatic rifle), and on and on and on. I mastered those weapons and was eager for combat simply because my brothers and so many of my friends were in combat, too. But God had his hand in it, somehow, because I came home unscathed. Actually, I got an assignment to return to the States very, very quickly, to prepare for the invasion of Japan. The Bomb had not been dropped yet…

So I went back to the States and was on thirty-day relief, and while I was on leave I went to see my brother in Baltimore. During my stay with him the Bomb was dropped on Hiroshima and Nagasaki, and we had a victory celebration, if you can feature such a thing.

But through all of the years, through the Civil Rights struggle, I was led to reflect on my war experiences and to come, slowly, to a position against war. That took the form of opposition to the Vietnam War. It all started in the early sixties – I was teaching at a black Catholic high school in New Orleans, and then we had the Cuban missile crisis. That was a turning point for me. Kennedy and Krushchev were threatening the very people I was trying to serve. They were playing God with my life and with the lives of people all over the South. That was a real crisis. We came very close to a catastrophe there; all of the southern Gulf ports were targeted by those intermediate-range ballistic missiles, and millions of Americans could have died. Of course, we would have slaughtered the Soviet Union in retaliation…

When I was in New Orleans I came to the realization that we were one of the most racist nations in the world. Hitler actually had read about

American racism when he was developing his sick philosophies. Then the Vietnam War came and I noticed that a disproportionate number of the troops going to Vietnam were black. And later on, after 1965, when I was stationed at a black parish on Baltimore's west side, I had to bury them when they came home in tin boxes. They accounted for a third of the combat troops in Vietnam but only one-tenth of the general population. These guys were mercenaries, fighting the empire's wars, and they had no rights as citizens back home...

THROUGHOUT U.S. HISTORY Americans seemingly have clung, with what can only be termed as religious fervor, to the belief that our nation possesses some inherent moral superiority which enables us to see the most clearly, which imbues us with the special power and authority to police the world. Nowhere is this reasoning more pronounced, and generally felt to be more indisputable, than in the history of America's response to the Third Reich. The development of National Socialism in the 1930s was a shocking and undeniable evil, and thus the Allied victory that cost thousands of American lives is rarely questioned as anything more than the triumph of democracy and goodwill.

Yet the United States has easily matched if not surpassed the Third Reich with regard to its stated goals of racial extermination. Native American people and culture were all but annihilated by the relentless application of belief in Manifest Destiny.

The term "Manifest Destiny" was coined in the 1840s to justify the westward expansion of Europeans across the American continent. What was said less emphatically was that in

order for Anglo-Saxons to accomplish this lofty goal, the indigenous populations would have to be eliminated.

This reality had not escaped General W. T. Sherman, and in 1866 he wrote to General Ulysses Grant: "We must act with vindictive earnestness against the Sioux, even to their extermination – men, women, and children."[14]

The pattern that began when the first Europeans set foot on American shores has changed only to suit new circumstances and to demonize new enemies. The thread running through each and every "conflict" is simply that we, the conquering power, are superior to the peoples we wish to subjugate and oppress.

Unraveling the myth of superiority is crucial to understanding the truth about war and the military, because it is the unspoken consent to myth that has caused fearsome numbers of young soldiers to die needlessly on the battlefield.

Retired Rear Admiral Eugene Carroll, the spicy ex-officer who once served on aircraft carriers in the Pacific, tackles this myth every day. When I interviewed him in his small office at Washington's Center for Defense Information, he minced no words. He is well qualified to criticize: in his lifetime, he commanded the USS Midway in Vietnam, directed Task Force 60 in the Mediterranean, and served on General Alexander Haig's staff in NATO from 1977 to 1979. He spent his last assignment on active duty in the Pentagon as Assistant Deputy Chief of Naval Operation for Plans, Policy, and Operations, in which capacity he was engaged in U.S. naval planning for conventional and nuclear war:

In the late seventies, I worked for General Haig in Europe. As part of my duties I was responsible for the security, readiness, and control of the use of 7,000 nuclear weapons. I had all the plans. I came back just absolutely convinced that this is madness. I got a job in the Pentagon where I thought I was going to be able to do something about it, and the second I expressed any questions or reservations about moving further out on that scale, I was confronted. I said, "This is foolishness. We're spending national treasure on something we could never use." "Well, you're nutty," they said. "You're not part of the team." And I wasn't.

I put in my request and left the Navy to come here to the Center for Defense Information to at least be free to express these opinions and to try to communicate knowledge and understanding to the American public, the press, the Congress, from a professional point of view. And that's what CDI does: we attempt to educate people as to what is wrong with our military strategy, our programs, the waste and excesses and the dangers of what we are doing...

We need to go out and tell people: we've got our own problems. The environment, drugs, education, racism, you name it. We've got things we should be working on at home, and you need leadership and you need money to do that. Instead, we devote money and great leadership, potential leadership, to preparing for the next war.

WHEN THE STORY OF MY LAI finally broke in 1969, the distinguished Protestant theologian Reinhold Niebuhr wrote: "I think there is a good deal of evidence that we thought all along that we were a redeemer nation. There was a lot of illusion in our national history. Now it is about to be shattered."[15]

British authors Michael Bilton and Kevin Sim, in their book *Four Hours in My Lai,* offer a similar conclusion:

The massacre at My Lai and its subsequent cover-up stand in the history of the Vietnam War at the point where deception and self-deception converged. If the Tet Offensive of 1968 had mocked America's complacent expectation of an imminent victory, My Lai's exposure late in 1969 poisoned the idea that the war was a moral enterprise. The implications were too clear to escape. The parallels with other infamous massacres were too telling and too painful. My Lai had been on the same scale as the World War II atrocities at Oradour in France, and Lidice in Czechoslovakia, outrages which had helped diabolize the Nazis. Reports now suggested that, if anything, the behavior of the American troops had been even worse. Americans, who at Nuremberg had played a great part in creating the judicial machinery which had brought the Nazi monsters to book, now had to deal with a monstrosity of their own making.[16]

Bilton and Sim go on to point out that "atrocity has been as much a part of the American experience of war as of any other nation...The circumstances may change from conflict to conflict, but the experience of atrocity is remarkably consistent."[17] To illustrate their point, the authors cite several examples from World War II:

The May 22, 1944, edition of *Life* magazine published a full-page picture of a "conspicuously decent and middle-class" girl writing a thank-you letter to her sailor boyfriend who had thoughtfully sent her the Japanese skull resting on the desk before her.

In April 1943, the *Baltimore Sun* ran a story about a local mother who had petitioned the authorities to permit her son to send her an ear he had cut off a Japanese soldier in the South Pacific.

On August 9, 1944, President Roosevelt announced that he had declined to accept the gift, sent to him by a serviceman in the Pacific, of a letter opener fashioned from a Japanese thighbone.[18]

The truth is that Americans are no better than any other people or nation. We are guilty of the same atrocities, the same oppression, the same injustice. We are not the global redeemers we believe ourselves to be, and may be most in need of redemption among all the nations of the earth.

BUDDHIST MONK AND TEACHER Thich Nhat Hanh (Thây, as he is affectionately called), was born in 1926 in central Vietnam. He became a Zen Buddhist monk at the age of sixteen and later cofounded the most influential center of Buddhist studies in South Vietnam, the An Quang Pagoda.

In the 1960s, Thây became the natural leader of the nonviolent protest against the "American War," as it was called in Vietnam, and was soon sent to the United States to seek the support of the American peace movement.

His work was not appreciated by Saigon, and in 1966 he was exiled from Vietnam. But nearly three decades later, Thây was still speaking out against war – this time, the war in the Gulf:

Immediately after ordering the ground attack on Iraq, in February 1991, President Bush addressed his nation, saying, "Whatever you are doing at this moment, please stop and pray for our soldiers in the Gulf. God Bless the United States of America." I suspect that at the same moment many Moslems were also praying to their God to protect Iraq and the Iraqi soldiers. How could God know which nation to support? Many people pray to God because they want God to fulfill some of their needs. If they want to have a picnic, they may ask God for a clear, sunny day. At the same time, farmers who need more rain might pray for the opposite. If the weather is clear, the picnickers may say, "God is on our side; he answered our prayers." But if it rains, the farmers will say that God heard their prayers. This is the way we usually pray...[19]

If President Bush had seen clearly the suffering of the Iraqi people, he would not have allowed his anger to be expressed by starting a war. He asked the American people to pray for the allied soldiers. He asked God to bless the United States of America. He did not say that we should pray for the civilians in Iraq or even the people of Kuwait. He wanted God to be on the side of America...

To those of us who are not American, this was not a good image of America. It was selfish and arrogant, and this was also a casualty that America suffered – not just by guns and bombs, but by your President's statement. If the President had said, "God bless us so that the war will end soon and that Americans as well as Iraqis will suffer as little as possible," he could have won a lot more sympathy from people around the world. But he did not say that...

We have to see how deep the wounds of war are. How could anyone call the Gulf War a victory? A victory for whom?[20]

We have experienced

all these terrible things, and now we're back in a society where everybody is running around buying each other Christmas presents, where life is "normal," and we can't even talk to them... What are we doing? What is all this about? *What the fuck are we doing?*

Brian

Brian Willson lost both legs, his left ear, and a chunk of his skull on Tuesday, September 1, 1987, on a naval base in northern California.

I read about it ten years later and decided then and there that I had to meet him in person. In November 1997, as I drove up to his house in Wendell Depot, Massachusetts, I was surprised to see him walk out onto the back porch to greet me. If I hadn't seen the dark wood grain of the prostheses above his sneakers, I might never have known about his legs. They had done a good job with his head and ear, too – he grabbed his left ear and wiggled it as proof. He was tall, with gray hair hanging down in locks to his shoulders, a red bandanna over his forehead. He smiled from time to time as he told his story.

I was born in Geneva, New York, in the Finger Lakes region, and grew up in a pretty traditional working-class family. It was a small town, conservative, racist; my parents, like everyone else, were very religious and very Republican. This was right after World War II, and America was everything, it was paradise.

I went to a one-room school house and was valedictorian of the eighth grade. At commencement I had to give a speech, and it was on sportsmanship. And one of the features of a good sportsman, I said, was that he never gives up, even when physically handicapped. So that was kind of prophetic…

By the time I had finished high school I was contemplating being a professional baseball player. I even had a contract offer from St. Louis, but I wound up going to college instead. Then, shortly after I got into college, I thought I wanted to be an FBI agent. I was very anticommunist, and back then, during the Cold War, it was just understood that the single most serious threat to humanity, to society, was the communists.

Next I became a born-again Christian. I'm not a Christian now, but it was part of my journey. I ended up at Eastern Baptist College near Philadelphia, but by the time I graduated from college I didn't want to be a minister anymore, so I went on to law school. But I still remember getting up at chapel in my senior year – we had a chapel presentation about U.S. foreign policy – and saying that we should just bomb the Vietnamese into oblivion and end the spread of this godless communism once and for all. I got lots of cheers. I even campaigned around campus for [Republican presidential candidate] Barry Goldwater in 1964.

In January 1966, during my second year of law school, I was drafted. I had never been militarily oriented, but I was totally for the war. I really believed in that war. So I ended up in the Air Force for four years, from 1966 to 1970.

I was still very straight-laced, and I taught Sunday school while I was assigned at the Washington, DC, headquarters as a second lieutenant. I was a big white male, and I used to talk with my friends about going out to beat up hippies. Hippies, of course, were the scourge of the earth, the downfall of our great nation. But even then my feelings about race were beginning to change – I noticed that I was a little more empathetic for anybody that wasn't white.

I got my orders to go to Vietnam in 1968, but I was sent first for twelve more weeks of training at Fort Campbell, Kentucky. I was ordered to an assignment at an army base, and we trained in patrols, ambushes, and other kinds of intelligence work. I had six six-man fire teams and two three-man mortar units under my command. Then I had to go for bayonet training. The trainers, who were sergeants, put bayonets on our M-16s; we were to shout "Kill" as loud as we could, one hundred times, and plunge the bayonet into the dummy. I was pretty athletic, and I didn't want to be identified as weak or unable to do anything. But I had a pretty hard time of it, and when the twenty other guys screamed "Kill!" I hesitated just slightly. The sergeant was on me right away, shouting into my ear: "Did you hear that order?" I was sure I could get myself psyched up to do it, but I couldn't, and the next thing I knew the sergeant had kicked me in the calf so hard that my legs buckled. That was an early sign that there was another dimension to my being.

I wound up going through that entire twelve-week training without ever having to use a bayonet again, and I was sent to Vietnam as scheduled. I was sent to Binh Thuy, ninety miles south of Saigon in the Mekong Delta, and as soon as I arrived I had to help set up base-perimeter security. We had machine guns and mortars, and I had three jeeps under my command. Our job was to fortify security at the base, and we were specially trained, with extra weapons and higher firepower, antipersonnel radar, and starlight scopes that allowed us to see at night.

So there I was, in March 1969, working around the clock in the 117-degree heat, sleeping when I could. We were attacked fairly often. Then I was assigned to go into villages that had been bombed by the South Vietnamese. They wanted a "gringo" to go around with one of the Vietnamese commanders, because they suspected that the bombers weren't following orders. I was only a rookie, what the hell did I know? But I went along. I thought it would be an interesting thing to do.

I'll never forget the first village we went into, just fifteen minutes after it had been bombed. We went in my jeep. What we saw was a lot of dead people, mostly mothers and children, and water buffalo. At least half of the villagers were dead; the others were wounded, moaning and groaning. The water buffalo that were still alive were roaring in pain. I came upon a woman lying on the ground with two children in one arm and one in the other. Her eyes were wide open, and it seemed that she was looking right at me. I bent over. It looked like she was dead. She was bleeding heavily, and then I saw that her eyelids were burned off because of napalm. The skin had just melted off her face.

In that moment – and it only took a second – I got it. I thought, is this what it all comes down to? Is this what my arrogance comes down to? I started crying. I stood up, and the Vietnamese lieutenant was looking at me: "Why are you crying?" I thought it was very obvious. I said something like, "Well, she could be my sister, and these kids could be my kids." He laughed – they were just communists. And then I knew I had moved on to something completely different. It seemed as though I now had a new soul.

(It's hard for me to talk about it, twenty-eight years later. It's really hard. Just recently I was at the VA hospital, screaming at the shrink, because I *had* to talk about it. I just sat there saying, "Fuck you, fuck you, fuck you!" Why do I have to bare my soul to *you?* There are a lot of feelings in there, and sometimes they just pour out like a flood…)

It took over an hour to get back to the base. We had to cross the Ba Sac River on a ferry, and I didn't say a word the whole time. I drove, and the Vietnamese lieutenant sat next to me. I went on one other such mission, and then I quit. I just refused to go into another village. I started speaking out against the war, and I had a pretty important position at the base. I stayed in the command bunker at Binh Thuy and tried to educate everyone about the senselessness of this bombing – how it was alienating us from the Vietnamese and needlessly killing thousands of civilians. I started studying the history of the war by going to the little base library, and I started, for the first time, to delve into history, genuine history. And it all started making sense to me.

Then one day I had to review a 7th Air Force report about the bombing of a village one month earlier. I suddenly realized that this was the same village I had gone into, and I was horrified. The summary said "130 VC killed." I knew that was not true – there hadn't been a soldier in the village. It had all been young mothers and children, and I was just outraged. Of course I was no longer an asset to the U.S. military, and I was sent home after only 120 days.

When I came back to the States, I still had a year left to do. Already in Vietnam I was told that I faced twenty years in prison for sedition, which was just short of treason. But I was assigned to a base in Louisiana, and the first thing I did was to sell my Corvette and buy a little Volkswagen beetle and put flower decals all over it. Big flowers. And when I drove up to the base to report for assignment, they stopped me at the main gate and said that flowers were considered political paraphernalia and were not acceptable. In fact they were supposed to be saluting me – I still had my blue officer's sticker on the bumper, and they were supposed to salute officers when they drove through. But I didn't give a shit about that. In the end, they wanted me to scrape the decals off, but I couldn't have done that without scraping off all the paint. So the chief of security said, "Well, we're just not going to salute when you come through the gate." And of course that was just fine with me, too. Already in Vietnam we had dispensed with saluting in my unit.

So I got through that whole year. I did a lot of antiwar speaking on the weekends, though not in uniform. I joined a Unitarian fellowship, and I gave a series of talks on the history of U.S. involvement in Vietnam, which lasted for five Sundays. A lot of people in the community came to that. The Air Force was also very interested in my talks, and they sent people there to monitor me. I'd see them in the audience and wave, and say, "Hey, I'm just exercising my First Amendment rights." I thought that if I had to do twenty years in prison anyway, I might as well keep on talking. What did I have to lose? My wife was concerned, but she was a lawyer, and by that time she was also against the war.

I finished my four years in the Air Force, went back to law school, and passed the bar in Washington, DC, in '72. I had become very radical, politically. I could now see that Vietnam was more than just a mistake, an aberration. It was part of our national ethos of arrogance, a predictable result of a long history of imperialism.

At this time I was ready to start my own practice. I had a master's degree in criminology. But I couldn't stand up with everyone else when the judge walked into the court room. It was like with the bayonet. I wanted to stand up, because I knew it was what I was supposed to do, but I couldn't. I also hadn't realized how much the flag hanging there affected me. As a boy I had carried that flag in my hometown parades – I had been so proud to carry that flag, but now it represented something very hypocritical. I remember thinking how I could ever explain these dilemmas to my wife. "It's a little, trivial thing, Brian – it's not a big deal. As an adult you have to adjust, adapt. You have to do these things if you want to be a lawyer..." But for me it *was* a big deal, and I decided that I wasn't going to subject myself to it.

I was almost ashamed of myself, but these things were emotionally loaded. In the end I didn't practice law. I couldn't handle the protocol, and I wasn't interested in business law or estates or wills or trusts or domestic law. So for four years I directed a program for the Unitarians called the National Moratorium on Prison Construction. I organized communities all over the United States to educate the public as to why more prisons were not going to be helpful for the future of our society; that we needed to look at alternatives to prison and at the basic socioeconomic reasons for crime. I lobbied Congress a lot, advocating against the Federal Bureau of Prisons budget started by Nixon. He wanted to build eighty new federal prisons. I ended up in Washington for ten years altogether...

In the eighties, I lived in Boston working as a legislative aide for a state senator. One of the things I did was to investigate inmates' complaints at a notorious prison in Massachusetts. In the course of thirteen months I investigated over four hundred brutality complaints and six homicides. Then, near the end of the investigation, I witnessed the beating of a prisoner in a cell block by two guards, and I had a flashback to the woman on the ground with the three kids, whom I hadn't thought about for more than ten years. He was screaming and they were stomping on him. At that point I had to get out of the prison. I couldn't even begin to address what I had seen happening. I just had to leave, to get out – I was totally shaken. I had thought Vietnam was behind me.

That was in June '81, and that's when I became a Vietnam veteran. Before then I hadn't thought of myself as a veteran. I was ex-military, but I wasn't a "veteran." I had been to Vietnam, but it was behind me. But that was a myth, and I realized at that point that I *was* a Vietnam veteran, and there was nothing I could do to change it.

That was when I took a leave of absence from my job. I had already been getting death threats at my apartment in Cambridge, telephone calls in the middle of the night: "If you know what's good for you, you'll lay off this investigation."

So I became the director of a Vietnam Veterans Outreach Center here in Massachusetts, filing claims with the VA, working with vets who were struggling with homelessness, drug addiction, alcoholism. Twelve Vietnam veterans committed suicide while I was there. That was no small thing to deal with. Fifteen, twenty years later, this war is more than a war, really. It is profound. It has pierced the bubble, destroyed the myth of the American way of life. There was something so fundamentally sick about Vietnam…

Then I started hearing accounts about what was happening in El Salvador and Nicaragua and Guatemala in the early 1980s. I heard things from people who had been there, and as I heard and read more in the regular news media, I started to get really upset in my chest and in my stomach. Whenever I read the words Marxist-Leninist, or Marxist-Leninist Sandinista, or Marxist-Leninist guerrilla, I saw the pattern. "Marxist-Leninist" was a code word that meant, send down money, send down arms, and kill 'em. And that brought up incredible feelings of outrage, grief, sadness, and anger. We were going to kill mothers and children again. But who gave a shit? We went on with our jobs, living our lives, buying newspapers that say whatever the government tells them to say. I started to feel the pain of the people, and once again I began thinking of that mother and her three children. It was happening again. I was hurting. And I couldn't remove myself from that connection.

By then I think I had a basic understanding of U.S. foreign policy, which was perhaps simplistic but ultimately true. I knew that the United States has less than 5 percent of the world's population and yet consumes anywhere from one-third to two-thirds of its resources. And to live like that you have to steal resources from around the world. We are imperialistic because we insist on living at a certain level of consumption, a level that is impossible to support unless we steal, collectively. That was a difficult reality to contend with. I didn't know how to deal with it myself. I knew that this realization pointed toward community, but I'm kind of dysfunctional myself, and I don't know if I could live communally. But I wanted at least to address this issue directly, even if I didn't know how to solve it or correct it. I wanted to be part of a discussion about it.

I started going to Nicaragua and El Salvador and Guatemala, and I became immediately aware of our policies by virtue of visits to many villages. I went to Nicaragua twenty times, as well as to El Salvador, Honduras, Guatemala, Panama, and Costa Rica. I spent a lot of time with the *campesinos,* the rural poor, and I just let them know that I was an American but that I could never support my government. The devastation was unbelievable. Our government's policy was torture, assassination, wiping out entire villages, destroying schools, blowing up bridges. Who can possibly support that kind of behavior? It doesn't take any intelligence at all to realize that this is not acceptable behavior. And it was my country against these people.

There were many other people, many other Americans who were also concerned, who were working in the Solidarity Movement. And in 1986 I joined with three other vets in a water-only fast on the steps of the Capitol in Washington, DC. We called it the Veterans' Fast for Life, and our goal was to change U.S. policy in El Salvador and Nicaragua. We wanted to encourage the American people to resist this murder, this destruction, and we wanted to make it clear that we were willing to die for this cause, if necessary. We had been part of the killing, and we were sick of it. We felt our country was in a state of emergency, inwardly, and we wanted to address the nation as war veterans and put our lives on the line. We had been challenged to put our lives on the line for war, and now we were putting our lives on the line for peace.

People were very upset with us because we said we were willing to die. That to them was a terrible thing. But we said, "Well, if you feel that strongly, why don't you show some concern for all the people being killed down in Latin America. They are dying, too, even if you can't see their bodies." We wanted people to make the connection between their concern for us, on the steps of the Capitol, and the women and children dying every day in Central America.

We sat there. We made wills. We knew where we wanted our ashes sprinkled. I wanted mine to be sprinkled in Nicaragua, in a little town up in the mountains. We had no money. We had no assets. But we were going to sit there indefinitely, until something happened or until we died. It turned out that after forty-seven days we were able to stop fasting. By that time we had received 30,000 letters, and 25,000 people had walked by us at the steps. We had been visited by 2000 people in a single day, and we had basically taken over the whole east side of the Capitol. That was a very important experience for me, because I learned then that I could not have any fear of dying. I had nothing, but I felt very free. I didn't have a job; I didn't have money; I didn't have life or health insurance; I didn't have a house; I had nothing to maintain. I was just there.

After that, I wound up organizing something called the Veterans Peace Action Teams. We went to the war zones of Nicaragua, and we rebuilt what the U.S.-backed contras had destroyed. During the Reagan years we had a motto: "What the U.S.-financed contras destroy, U.S. veterans will rebuild."

So we went into these villages and rebuilt health clinics and schools. It was an atonement, a sign that we were no longer part of the violence. We could not be silent; we had to be active in some way, in a personal way. And we ended up sending eight teams to Nicaragua.

While all of that was going on, I got involved with the Nuremberg Actions in California. These were a series of actions aimed at blocking the munitions transports then being sent from the Concord Naval Weapons Station to El Salvador and Nicaragua. That's where I got hit by the train. Actually, it was all very well rehearsed – we were in many ways a fairly conservative group of people. At least, we never did anything rashly. We thought about everything we were going to do very carefully; we even notified the government about our planned actions. Being a lawyer, I studied the history of train blockades – I prepared myself for two months before blocking that train. I knew it was going to be a hard experience, a fast of forty days, just on water, sitting in front of the death trains. And I figured that the worst that could happen would be jail. I figured I would be hauled off to jail once the fast had weakened me.

The rest is history. The train did not stop. Fellow veterans Duncan Murphy and David Duncombe managed to escape with minor injuries, but Brian, seated in the middle of the tracks, was run over. The train dragged him twenty-five feet. Its operators had received orders to drive through the crowd to prevent anyone from boarding, and they increased the train's speed to three times the legal limit. The Navy meant to make their point, brutally and unambiguously. As Brian lay there bleeding to death – both legs severed and a hole in his head gushing – Navy personnel stood by, denying medical assistance and transportation to a hospital. Brian was not on Navy property, they said.

It was only thanks to the medical knowledge of Holley Raven, his new wife, a trained midwife, and Duncan, a World War II medic, that Brian was kept alive for twenty-three minutes until an ambulance arrived. After nine hours of surgery, his ear was sewn back on and his skull patched up, but it took him days to realize that his legs were gone.

I don't even know how I survived. I lost both legs below the knees, lost a big piece of my skull – a piece of bone from my cranium went in and destroyed my right frontal lobe. Most of my ribs were broken, my arm was broken, my shoulder was broken, my ear was cut off completely. My right kidney was basically destroyed. Hey, I have been given the gift of a second life!

I still don't remember being hit by the train. And they say it took another five days before I was really conscious. I remember waking up, looking around the room, and then noticing all these green plants. I thought, what an interesting jail cell – I assumed that I was in a prison. My wife was there, and she said, "Well, Honey, the train ran over you. It didn't stop." I thought she was kidding. And then I started to get mail. The day I was hit on the tracks I had no money, no health insurance. But I got thousands of letters every day, and within two weeks 3000 people had contributed $200,000 to help with the hospital bills – totally unsolicited.

For the first week or so I was just happy to be alive, like, "Wow! How could I have survived?" And then I remember feeling that I was now somehow part of a new peer group. In the spring of 1987, just months before this all happened, I had been in Nicaragua. I had been at the bedsides of more than four hundred amputees, all victims of our American mines. After shaking each man's hand I had told them how much they motivated me to do even more work back home.

I thought about hating the train crew, the people who'd run me over. But I knew that hatred would destroy me – I've seen what it does to people. It destroys their bodies, their minds, their spirits. So I had to work on being liberated from hatred.

I also thought about my parents. They were crushed. Here I was, the only kid in my town to get a real education. The son who was supposed to be a success, make lots of money. A military officer and a lawyer with a master's degree and a Ph.D., sitting on the railroad tracks like a bum. Fasting on the steps of the Capitol and flying off to Nicaragua. With no money and no health insurance. They were devastated. It was very, very hard for them to comprehend, and I could understand why.

Already many years earlier, my father had disinherited me to [right-wing televangelist] Jerry Falwell. In his will, he wrote "Jerry Falwell" in lieu of me. He was a big contributor to Falwell, to the Moral Majority. And Falwell contributed a lot of big money to the contras. My father got this newsletter from Falwell describing how Reagan's freedom fighters, the contras, were liberating Nicaragua from the Marxist-Leninist Sandinistas. And he would tell me: "Falwell has a Ph.D. He's a man of God. And you have renounced Christianity, and you are supporting the communists." I would argue that I really wasn't supporting communism but rather condemning imperialism, and that I had a Ph.D. too, and that I was also his son. But my father just couldn't make it out. That was really painful for me.

I had hated my father already, for years. That's one of the reasons I know hate is so damaging, because I hated my old man. And he really was a mean, cruel father. I know from the men's groups I've been in that this is not atypical. In fact, it's almost universal. It's fairly rare among men in my generation to have had fathers that were emotionally or even physically present enough for us to know that they really loved us. Hardly any of us have had fathers who showed anything that could have been interpreted as love. Of course, they were simply imparting to us what they knew; I don't think they ever intended to screw us up. So it was pretty hard, but I made peace with my father in 1983, and by the time my father died, in

1989, I actually loved my father. Before he died he even changed his will – took Falwell out and put me back in. And I had nothing to do with it.

I made peace with my father by realizing that I didn't have to like him to love him. What a concept: unconditional love. Doesn't have anything to do with liking a person. It's nice if you like him, but it's not necessary. I had to find a whole different attitude toward my father, because he wasn't going to change for me to love him. I had to change. I had to learn to love somebody that I didn't like. That was a big help to me later, with the train crew. Obviously I was devastated by what they did to me, but I can still have a sort of love for them.

I guess some people can go and do acts of civil disobedience without even thinking about it, but I have to think about it, because I don't want to think of the police as pigs. I have to prepare myself, and I'll only do it when I feel that I can present a model of love. I have to spend time talking about it with people, get very centered, do a lot of breathing. When I went out to confront the military on September 1, 1987, I wanted to be loving. I had anticipated that to be loving would be very difficult: we were going to be pulled off the tracks, probably gruffly. And having been a right-winger myself, I could imagine what I would feel like if I had to deal with the "goddamn stupid-ass people sitting on the goddamn tracks, what the hell do they think they're doing" demonstrators. I understood that way of thinking.

So I never went through a period of hatred. I went through a period of grieving, but not until seven years after I lost my legs. That was in '94, and I went through months when I cried every day. I was crying for more than my legs: the loss of my veteran buddies, my parents' death, my second divorce – I was suddenly dealing with all these losses at once. It went on so long that I began to think I'd never get over it, but it turned out to be very cathartic. I became much more deeply grounded after that.

I do a lot of men's work now, trying to get men to give up their definition of manhood as machismo – to start being emotionally honest, to leave the competitive world, to get in touch with their feelings, to stop pretending that they're tough and disinterested and always in charge. I'm after a transformation of some kind. I don't care how you define the meaning of life spiritually or religiously, but I am concerned about becoming aware of what is sacred. In fact, everything in life is sacred, but the concept of sacredness has been lost in our post-agricultural era. We've been seduced by technology and materialism and science, but if we are to find meaning in life again, we have to search for what is sacred.

We vets were simply the right age at the wrong time. And our personal identities have been shattered, our loyalty to the nation-state has been shattered; our religion, our psyches and egos have all been shattered. We came back to a society that couldn't deal with us. The war was much more than a lie – it was demonic, diabolical. And I am convinced, after working with vets, that we need to find new souls, a new meaning for our lives, which is completely different from what we have believed in before. And how do you do that? Who do we talk to about it? Who do we share that process with? All these guys are blowing their heads off with shotguns. Their pain is that intense.

Who am I? What is the meaning of life? We have experienced all these terrible things, and now we're back in a society where everybody is running around buying each other Christmas presents, where life is "normal," and we can't even talk to them...What are we doing? What is all this about? *What the fuck are we doing?*

I don't think people grasp how profound the Vietnam War is for the United States of America – not only for vets, but for the whole society. To understand the Vietnam War, you first have to try to understand our society. And you begin to wonder, "Gee, maybe our whole society needs to go through a healing," and no empire has ever done that. We vets, of course, have no alternative. We have to choose a journey of healing and transformation or else check out. And even if society doesn't seek this healing with us, as individuals we have to find healing if we want to survive. We have to get really clear about our own personal lives if we are going to navigate ourselves successfully through this culture, which, in many ways, is a very sick culture.

11 The Road to Peace

If only there were evil people somewhere, insidiously committing evil deeds, and it were necessary only to separate them from the rest of us and destroy them. But the line dividing good and evil cuts through the heart of every human being. And who is willing to destroy a piece of his own heart?

Alexander Solzhenitsyn

In October 1997, I attended a weeklong retreat for veterans held at the Omega Institute in New York's Hudson Valley. Zen master Thich Nhat Hanh and a small group of Buddhist monks and nuns led the retreat. During the week, many veterans shared their stories by writing them down and then reading them aloud to the whole group. This process of opening one's pain and burden to the *sangha,* or gathered community, was powerful and cathartic.

Aside from writing, we also talked a lot – when we didn't feel like writing and needed others to listen, or needed to listen to others.

Marshall, a middle-aged man, had a story which seemed too horrible to tell. He'd worked for the CIA in Operation Evergreen. "I thought I was doing the right thing at the time," he said, choking. "But I remember this six-year-old girl…" He stopped, unable to say more. He often had to leave us, once placing a note near his pillow on the floor that said simply, "God bless. I love you all. Have to go process." Later, he told us that he dreams about meeting the people whose lives he has taken. In the dream, they are alive; he says he is sorry, and they embrace and laugh. But then he stops short: "I don't deserve forgiveness."

Nikko has wept rivers of tears. As a young man, he tried to resist the war, refused the draft, and joined antiwar protests at school. But he was inducted only months later and was sent to Vietnam as an Army medic. He was there at "Hamburger Hill" – one of eleven men to survive from a company of 155 GIs. His company was sent up that hill twice a day for twelve days. His comrades recommended him for the Medal of Honor,

but the government didn't even give him that small satisfaction. They told him that a conscientious objector who is still alive doesn't get the Medal of Honor. Nikko came back an old man, his hair turned white from the stress of fear and death.

We were touched, too, by Phâp Úng, a Vietnamese monk, who came in unannounced and left a short note with the retreat's facilitator, explaining that he wanted to share our pain and help to alleviate our suffering. Bob, a Hispanic vet, said he felt resentment at the presence of the Vietnamese – they forced him to bring to the surface things he thought he'd put behind. Ron Orem, dark-haired and boyish-faced, echoed this sentiment, but acknowledged that such confrontation was a necessary step to reconciliation:

When I first saw your group of monks and nuns coming toward me down the path, I thought, Oh, shit! It's the little people again! I didn't know if I was supposed to jump into the bushes or open up or what.

Every time I see you, I am reminded...I want to go back to your country one day without fear, to sleep in your jungles without claymores and trip-wires. I have to deal with you because you are here. And I know I have to meet you face to face much more.

We broke up into smaller writing groups shortly afterward, but Ron and the young Vietnamese monk knelt face to face, completely absorbed in listening to each other. Ron spoke softly to the monk – I couldn't make out the words, but I didn't need to – over his shoulder I could see the tears rolling down Phâp Úng's face, his eyes locked on Ron's. I was trying to concentrate on writing but looked up when I heard what I thought were the sobs of a child. Phâp Úng and Ron

were embracing one another. They held one another for what seemed an eternity. Ron held the "little people" and the monk held the soldier. There was forgiveness and peace.

At last they let go and wiped away the tears. Everyone felt their joy. Later, Ron read something to the group: "I used to hunt people like you; now you hand me a flower. I am confused…"

JUST WEEKS AFTER the retreat, I received a letter from Nikko, the survivor of "Hamburger Hill." His reflections exemplify the type of healing that veterans of all wars must – and can – find:

The impact of our time together has led me to a whole new world of insight and healing. Since leaving Omega, I have celebrated Veteran's Day –a first – by being with other vets, and reading some of my writings with the local Buddhist group in the San Francisco area. The result has been a sense of peace around my Vietnam experience that I've never known before. It is as if I am finally beginning to put that part of my life into its proper place in my own personal history, and I no longer feel like I am toting around a huge bag of smelly garbage that I am incapable of letting go of or dealing with…

I am even sensing a need to become involved in socially conscious programs again. That's a huge step for me, as I haven't had the energy for that in twenty-five years. I am still in pain, but thanks to our being together I have begun to move on from the place in which I have found myself stuck for so many years. It seems to me that if, in the course of our stay on this earth, we can impact even one other person in a significant way – then that is a powerful reason for "being."

Clearly, one of the most effective catalysts for healing involves creating "spaces" in which veterans can speak about their ex-

periences and emotions in a group setting. Again and again I have heard statements like "nobody else knew I was a Vietnam vet," or "for twenty years I never told anyone." But just as often I have heard veterans say that until they could talk openly about their experiences, the road to healing was blocked. However, they stress the need for an audience that listens without judging, that seeks to share pain rather than to point fingers.

Such opportunities demand time, care, and a willingness to listen. The benefits of simply having others to talk to cannot be underestimated. But therapy groups can become ends in themselves and may cause more harm than good. In an atmosphere of such overwhelming need and suffering there is the danger of seeing oneself either as a victim, hurt beyond forgiving, or as an inhuman monster destined to wallow forever in guilt. The two may seem far apart; in reality they are close twins. In the intensity of the struggle, the pull to either side is so strong that the survivor is left teetering on the brink.

When Vince Bramley returned to Aldershot after the Falklands War, he was welcomed as a hero. But seventeen years later he still cannot shake the memories of his encounters with death on the mountains of East Falkland island:

The feeling of shock has stayed with me for sixteen years; it's still with me now. That's never going to go – I'll always feel the same punch in the gut. And it's changed me. I mean, before the war I was known to be aggressive, always spoiling for a fight. But I've not been in a fist fight for the last ten years. I've had arguments, disagreements, but I've always walked away. In the same context, I don't think that I have to prove to myself that I'm a rough, tough geezer with an image of "I've been to war, I've killed people." Because that doesn't make you rough and tough at all.

Within eighteen months of the war's end, something like sixty to sixty-five percent of those blokes in 3 Para had left the Army. I think that speaks volumes in its own way.

I used to have nightmares at least once a month for sixteen years. Then in 1997 I returned to the Falklands with a friend of mine, Dominic Gray, who was shot in the head during the war. (It was a billion-to-one shot – went right through his head and came out the other side. We always say it made him madder, but he's perfectly sane: he's earned a law degree since.) He and I went down there to work for a civilian company at Mount Pleasant, where they've built a new airport, and we helped install the fire and smoke alarms throughout the whole camp. It gave us a unique opportunity to return to the Falklands and see the place sixteen years after the fighting.

Dom and I went back to Mount Longdon and retraced our own particular battle steps. Then we stayed up on the mountain for a night, holing up in an old Argentine bunker. We'd brought some food, and we just sat up there and had a little chat on our own. In the morning, I laid a cross for one of my school friends who was killed beside me. I've been back in the U.K. sixteen months now, and I've not had one nightmare. Neither has Dom.

I guess you could say we buried our ghosts that night, sitting up there on Longdon on our own. Yes, I still get upset about some things, because there will always be things you don't like to remember. But you've got to remember them anyway, until the day you die, and I know that.

Over the last sixteen years, I've come to terms with the fact that I will always see Denzil Connick rolling around with his leg blown off (I was one of the first to reach him). And I will always see Craig Jones dying in my arms. He was only twenty-one, I think. I watched his life ebb away. He was smiling at me. And I've always wondered, did he think he was going to survive because we were there beside him, or was he smiling because he realized he was going and he was letting himself go in the arms of a friend, or at least a fellow soldier? I've always had sad, sad feelings about that, because I don't think he deserved to die. None of them did. But to see someone die literally, you know, smiling at you like that is – I've had a horrendous time with that for years…But I accept that I will always see those things. You can't block them out.

Like many other veterans, Vince has used writing as a means to process his feelings and to come to terms with a war that has changed him for life. After writing *Excursion to Hell,* his best-selling account of his own experiences in the Falklands, Vince began exploring the idea of a book that would examine the war from the perspective of soldiers from both sides. He traveled to Argentina and met and interviewed eight soldiers from the 7th Infantry Regiment, men he had tried to kill during the battle for Mount Longdon. Drawing on their accounts and those of five 3 Para veterans, Vince wrote *Two Sides of Hell,* a window into the battle they fought the night of June 11, 1982. Of the Argentinian veterans he says:

They're not just targets that fall anymore. I've had the opportunity to go and meet them face to face; I've been to their homes and shared meals with them. And laughter. And tears. Some of them are actually friends of mine, and we write to each other. I couldn't kill them now.

I went to Argentina because I wanted to know what the enemy was like, really. I wanted to know them as people and to educate myself about them. While researching my book, I sat across the table from people who had killed friends of mine, and I'd killed friends of theirs. We felt no hatred for each other, only respect.

I didn't want to write just another book about us winning the war. And I think that the way it turned out, when you read in the book about the battle for Mount Longdon, you don't want any of them to get hurt or killed, whether it's Jorge or Dom, Antonio or Denzil, or whoever: they're all human beings.

Fellow paratrooper Denzil Connick, who lost his leg to an Argentinian shell, accompanied Vince to Buenos Aires, a trip paid for by Argentina's *Agente* magazine. There they met face to face with soldiers of the 7th Infantry Regiment, men who had been their enemies during the battle for Mount Longdon. It was an encounter Denzil will never forget:

There were only six howitzers firing onto Mount Longdon from the town of Stanley, and I met a guy who was the commander of one of them. So there was a one in six chance that he was the guy who pulled the cord on the gun that blew my leg off.

When I met him, I felt no anger towards him. We were both too emotional to say an awful lot straight away. There were a lot of tears, and we just basically hugged each other. There was nothing else we could do. Later on, through interpreters, we talked about how we are now, about our own families, about how we felt during the fighting. This guy felt really sorry, now that he'd really actually met one of the people he might have hit. Being a gunner, he'd not really seen too many British casualties from his handiwork, if you like.

I'm not angry at the Argentinian soldiers; they were just doing their job, like we were. But I get very angry when I think how the British Army left me high and dry after they'd had their worth out of me. Today I'm absolutely no use to them whatsoever. I'm a lone wolf. I used to run with the pack, but now I am no good to the pack. They don't want you anymore, so you leave. You have to survive on your own. Without my family to lean on – and I mean lean on – I don't think I would have made it. Others haven't been so fortunate. A hell of a lot of guys have just killed themselves because they just felt so utterly worthless. They are no good for anything any more. They have been rejected by the military, rejected by the government they supported. It's terribly, terribly lonely to feel like that.

Jim Savage went to Vietnam at age eighteen, in April 1965; five months into his tour he was hospitalized for a severe infection and then malaria. After losing sixty-three pounds, he was transferred, still feverish and delirious, to a hospital in Japan. There he spent five weeks in isolation, during which time he learned that his wife had given birth to a premature child, which died.

Jim made it back, but many of his friends didn't. As we sat together, he reflected on his own journey to healing, one that could not start until he could begin to talk about what he had seen and done in Vietnam:

A lot of things stay with you for years and years – things you don't want to talk about, things you can't even discuss with your own wife. I used to wake up at night with hallucinations, with dreams and nightmares, for years after coming home…

During my tour of duty there was an incident – our perimeter was broken through, and we had a firefight. It's very hard, very difficult, to have to take someone's life. But it's even harder when you have to do it in a close situation. The memory stayed with me, very vividly, for years. The eyes, the facial expression – everything stays with you for a long, long time. It's something I don't think I can ever forget.

When I first came home I drank heavily and secluded myself, put myself in a shell that nobody could penetrate. I had trouble dealing with other people, and at work I insisted on being alone. I didn't want people around me. I had a lot of anger, a lot of hate, a lot of prejudice inside me.

I also had a lot of fear, which people can't understand if they haven't been through it themselves.

I was looking for forgiveness, from people or from God, trying to convince myself that there could be healing. I knew I had to face these things. I couldn't pretend they weren't there, because they would always raise their heads. Certain things would bring them out. You get overworked or overtaxed or overtired, and the next thing you know, your dreams are very vivid again.

It's like taking out the garbage. You can put your garbage in the can, but if you don't remove it, get it cleaned up, it stays right there. That's what I did. And I tried to ignore it, pretend it wasn't there, wish it away.

Over the past twenty years, Jim has slowly begun to find healing. For him, forgiveness was possible through a newfound Christian faith – forgiveness for himself, his family, the people around him who "couldn't understand," and the warmakers who had indirectly caused him so much pain and suffering. His most important breakthrough, however, was the realization that he would have to confront the monsters in his own closet before he could ever hope to break free from the nightmares of his past. "There were things I was still keeping way down inside, but that needed to be brought out," he said. "They weren't easy to look at, and I didn't want to do it, but little by little the walls came down. And I found I was able to be around other people, too. We used to say that some people came home, but never got off the boat. That period of your life has to be dealt with – it has to be talked about. And very often it's other veterans that can best help you to go on from there, because all of us have scars..."

Bruce Cole saw Vietnam in its full, bloody horror. He enlisted after getting his master's degree and ended up in the infantry, a stupidity that overwhelms him now: "When things are going good, I get scared. One, because I feel I don't deserve good things to happen to me. And two, because that's when ambushes happen – when things are going good. In Vietnam, I got away with murder…I keep waiting for some kind of revenge or retribution."

He told me about the first veterans' retreat he ever went on, and how it helped him to confront the demons of his past. He had tried to write down his experiences in Vietnam but had gotten stuck trying to explain how he had lost his first man. Until then, he couldn't remember just how it had happened, but the writing process triggered memories suppressed for decades by shock. Frightened, he told his wife, "I'm going to need your help. I think I know how it happened. O God, I remember now!" He wrote it down and showed it to her. They held one another and cried together under the trees. When they returned to the group, Cole wanted to share his story to be free, but he didn't have the courage to read it aloud. Finally, with another veteran holding one arm and his wife holding the other, he told his story, fully, for the first time.

Call it what you will – group therapy, cathartic release, getting in touch with your tears – it is very close to what Christians call remorse, confession, and repentance. There is great healing in sharing the darkest secrets of our lives with others. As the Buddhists say, "Releasing what is deep inside you will save you; keeping it to yourself will destroy you."

HEIDI BARUCH, A NURSE, was in Vietnam for two years. She doesn't look old in body, but her eyes say that she's seen it all, that she was long ago stripped of all her innocence.

I went for the first time at the age of twenty-one. I grew up in a town of three hundred people and had never really left home until I went into the military.

I came home after the first year and I didn't fit in anymore... I couldn't bear the thought of all those people dying back there, either, without the support of their countrymen back home. So I went back for a second tour. I worked primarily in the emergency room and in intensive care.

I don't remember being with many people when they died, but I know I was. I remember a number of more than a dozen in one night shift, and I only remember five specific deaths. But they have never held a name or a face, just a haunting memory. The first was in the ER when I didn't know how to triage yet. I tried to resuscitate a man when his brains fell out in my hands. I don't remember his face, but I remember going through his wallet, and then I saw he was a son, he was a father, he was a husband.

Then the next was the intensive care unit. That stupid doctor wouldn't let him die, but he was dead. His legs smelled of rotten tissue, so we kept putting blankets on him so the smell would be reduced. We would take turns sitting with him. When his blood pressure dropped we had to call the doctor, so he would make us give more medicine to raise his blood pressure and keep him breathing. And he wouldn't let him die. I finally remember screaming at the doctor to let him die. And I felt the guilt for having to have an argument at his side. I hope he knew I was on his side. I wanted him to have peace.

Then I remember the three. They burned up in a tank. The stupid government built those damn tanks so the guys had to crawl out over the gas tank when they exploded. They got out, but they were charred. We didn't know if they were white or black. They didn't.have an inch where we could even put an IV. They had no pain, except the pain of lying there waiting to die. And they died one at a time. And they would ask, "Am I going to die?" I couldn't say no. I could only be there.

It was always important for me to be there and not let someone die alone. I don't know why. That's all I had to offer. There was a time I started to write letters to families just to say, "I want you to know your loved one did not die alone." I guess death was so scary to me that I did not want them to be so scared when they died.

I didn't know how to deal with death, and I didn't know how to deal with it because my stupid religion didn't offer me any support. Nobody offered me support, because everyone else was in the same awful place. We only knew how to support each other by the stupid jokes, the alcohol, the sex. How could the government think so many mixed-up people could do anything with any effect?

The whole stupid war was a mess, and it seemed no one cared. That's why I had to go back, too, because no one cared. So I had to go, and I had to care. But how stupid I was, because I don't think I knew how to care. I went back a second time because it was the only place I fit in. I was a machine. Vietnam stripped me of all emotions...

In triage, I was told to keep Americans alive but let the Vietnamese die. Nobody should have to make such choices. I went as a girl of twenty-one, but I came back an old woman. I wanted to help, but I wasn't prepared to see what I saw. When you live in an environment of hate and anger, you become hate and anger...

You have to forgive in order to heal. A lot of Vietnam vets go around feeling victimized, but you have to turn it around to begin to heal. You have to start taking responsibility for what you did.

According to Vietnam veteran Steve Bentley, there are two basic factions among Vietnam veterans today: those still

"stuck" in the war, and those who have moved on. The difference? Steve thinks it has a lot do with whether or not a vet has actively sought opportunities for reconciliation.

Such an active search for healing and reconciliation has changed the life of Philip Salois, whom I met in May 1998 at New York City's Hunter College for a conference on the emotional wounds of veterans. Phil is the national chaplain for the Vietnam Veterans of America; he went to Vietnam in 1969 and served with the 199th Light Infantry Brigade. His baptism of fire came on March 1, 1970. During a search-and-destroy mission, his men had found a battalion of NVA, backed off, and camped the night, waiting till morning to attack. The next day the company commander foolishly ordered them back down the same trail, where they walked right into a U-shaped ambush. Six men were immediately separated from the rest of the group, and a long firefight ensued. Phil finally volunteered, along with two other men, to try and rescue those six men.

One of the volunteers took a round in his shoulder and was immediately incapacitated, but Phil and his buddy Herbert Klug made it out to an open clearing, where they laid down enough diversionary fire to allow four of the six men to escape. It was then that Phil realized Herb was no longer with him. He was lying face down in the grass, a bullet hole in his chin.

Phil had a "foxhole conversion" in Vietnam. "I told God that if he'd get me out of that mess without a scratch, I'd do anything he wanted," he says. But it wasn't until 1974, while in the second year of seminary in Los Angeles, that he remembered the vow he had made. Almost ten years later, in 1983, he decided to attend a conference called "Healing the Vietnam Veteran," thinking he had something to contribute. But by the time the conference was over, he realized that he himself had not yet dealt with his time in Vietnam. There were ghosts from his own past that needed to be confronted before he could help others.

He enrolled in a six-month program for vets with PTSD, hoping to deal with the tremendous guilt that he felt over his friend Herbert's death. Why had Herb died while he lived? The rescue had been his idea, and this added to his torment.

He also realized that he still had negative feelings about the Vietnamese people, that he harbored mistrust and contempt for them. Finally, in 1990, he was able to return to Vietnam, to the battlefields where he had once fought. He went to a cathedral in Saigon and confessed to the congregation that he was a former American soldier who had once fired on their relatives. He asked their forgiveness, and he found healing and reconciliation in their presence. But it was the children of Vietnam who reached out to him and touched his heart, and it was through them that he began to look beyond himself and toward his fellow veterans.

In 1991, Phil came to the realization that his healing had to include visiting the family of his dead buddy Herb Klug. He tracked them down and spoke with Herb's father. Then he wrote Mr. and Mrs. Klug a long letter, explaining who he was and what he wanted to do. In July 1991, the Klugs picked Phil up at the airport and took him into their home in Dayton, Ohio. The day after he arrived, Mr. Klug took him to Herb's grave.

Phil wanted to "talk" with Herb alone, but Mr. Klug wanted to stay, so Phil began his conversation with Mr. Klug at his side:

> Herb, I'm finally here. And I'm sorry it took twenty-one years to get here, but you can be assured that I have never once forgotten you or the sacrifice you made on March 1st. I have always included you in my work with other vets, and I believe that we're doing this work together. Hardly a day goes by when you are not included in my thoughts and prayers…Be at peace, brother, and stay with me!

When Phil stood up, Mr. Klug was sobbing. Phil apologized for opening old wounds, but Mr. Klug said, "No, it's not that at all. I listened to what you said, and it made me think that I never once took the time to tell Herb that I loved him. And I never talked to my wife about his death. I always felt I had to be strong for her."

Around the supper table that evening, Phil told the Klugs how their son died, and related the events leading up to his death. He told them how guilty he felt that Herb died while he himself lived, and that the rescue had been his idea. Herb's mother, Beulah, grabbed Phil in both arms and said, "Never say or think that again! We never held you responsible for the death of my son."

Phil's honesty allowed Beulah to speak of her own pain. She remembered how her Baptist minister had come up to her during the funeral and said, "Oh, Beulah, if you'd only come to church every Sunday, maybe God would have let you keep your son." She never set foot in church again.

On the twenty-fifth anniversary of Herb's death, Phil called the Klugs again. He wanted to organize a reunion for his platoon, which would include a cookout at the Klug's house and a ceremony at the cemetery. The Klugs thankfully agreed, and a sizable number of men gathered in Dayton at the Klug's home. One of them was among those whom Phil and Herb had rescued from the ambush, and he expressed deep gratitude to the Klugs for Herb's life, which had been spent to save others. Beulah was overcome with emotion. She could hardly believe that so many people still cared about her son after so many years.

"Healing is very hard work," Phil says. "In order to heal, you need to *do* something. You need to move. You need to be creative and find healing for the whole community." He also maintains that spirituality is an essential part of healing – that talk therapy and medication can only do so much. "We need to get in touch with the mystery of ourselves," he says. "I certainly couldn't face the demons, the ghosts of my past, without the Christ-person inside of me."

In collecting stories of how veterans are "dramatizing" the healing process, I have been amazed at the number of ongoing projects Vietnam veterans are involved in. Jack Meagher has dedicated himself to "PeaceTrees," a group that plants trees and creates parks, not only in the inner cities of Washington, Los Angeles, and New York, but also in the minefields of Southeast Asia. In Vietnam, they first had to clear the land of unexploded ordnance and landmines, and in the summer of 1996, they removed 300 landmines and 1500 pieces of mortar, bombs, and shrapnel to turn an 18-acre battlefield into a park.[1]

Nurse Heidi Baruch has spent the last six years treating veterans with severe PTSD (post-traumatic stress disorder) at a VA hospital. She said she used to think that she was helping veterans, but now she realizes it is they who are helping her to heal. Recently, she joined Vietnam Challenge, a 1200-mile bike trek from Hanoi to Ho Chi Minh City (formerly Saigon) undertaken by eighty American and Vietnamese military veterans in January 1998. The Challenge was not a race or competition, but a chance for the participants, who once fought against each other, to find emotional healing and a sense of closure. Many of the cyclists had physical disabilities, among them a young man born blind due to Agent Orange exposure who rode tandem with his veteran father. And at the end of the ride the U.S. contingent gave Vietnamese officials a check for $250,000 to help rebuild Bach Mai – once Southeast Asia's biggest teaching hospital, but never fully rebuilt after it was leveled by a B-52 strike in 1972.[2]

After returning to the United States, Heidi wrote:

I am at peace with Vietnam after thirty years. I spent twenty years pretending that Vietnam had no impact on my life, but I always knew that it had – my life *was* atypical, although perhaps no different from those others who experienced the war…

I have now received treatment for PTSD, and have been fortunate to return to Vietnam. I was able to visit three of the four places where I served as a very young nurse, and instead of the intense emotions I expected, I felt a wave of release and then peace. Something changed inside of me when I saw those places – when I saw cows grazing quietly where field hospitals once stood. I was just amazed at how green everything was. I'd always remembered Vietnam in "dark" colors – the colors of fear, hate, anxiety, and helplessness – but now I'll think of Vietnam as a beautiful, colorful place of peace. Every vet should return to Vietnam!

That recognition is an important part of healing is made clear by the hurt that results when it is denied. Richard Luttrell was haunted for three decades by a photograph of a little girl standing next to a man – the Vietnamese soldier he had killed – and when in 1997 he finally received a letter from the man's son, confirming his identity, he was ecstatic. His goal was one day to find the girl and apologize for what he had done. When her brother's letter arrived, it seemed that this wish might soon become reality.[3]

I joined the Army in 1966 and was assigned as an infantryman. Then I went through jump school in Fort Benning, Georgia, and from there I was assigned to the 82nd Airborne Division at Fort Bragg, North Carolina. I was there about a year before going to Vietnam, where I served with the 101st Airborne Division.

I was pretty much a naïve adolescent when it happened. We were on a mountain ridge, and we were staggered on the trail facing opposite directions, about fifteen yards apart, as a safety measure. The point man was teaching me to walk point, so I was walking slack. Then they held the column up – I think they wanted to get a radioman up close or something – and as we were standing there resting and taking a break, I lit up a cigarette. As I did, I saw, out of the corner of my right eye, an NVA soldier squatting and pointing a weapon at me on the trail just thirty feet away. We stared at each other for what seemed like an eternity, and then I fired and killed him.

There was return fire from both sides, and when it was all over, we moved up and found three dead NVA. We didn't have any casualties. I remember the guys going through the NVAs' personal belongings, taking belt buckles and belts – souvenirs, so to speak. One of the troopers offered me the belt buckle of the guy I had killed, and I said, "I don't want it." I was just sick. It was surreal. I couldn't believe it had really happened – that I'd actually taken a life. I'd never seen anybody shot before, so it was pretty traumatic.

One of the guys went through the wallet of the soldier I had shot and threw it to the ground. I was still standing there in shock when the command came, "Okay, let's move out." We just left them lying there, but as we started down the trail I looked down at the wallet. A picture had fallen out of it. I bent down and picked it up. I looked at the picture and looked at the soldier. It was obviously his picture; he was in uniform, and there was a little girl in the photo, too. Her hair was braided in pigtails. It looked like she was about eight or nine years old. I don't know why, but I took that photo, and I carried it in my wallet for the next twenty-two years.

I'll never forget the look in his eyes. I'm sure he could see it in my eyes, too. I was scared to death, although he looked more solemn and sober. I found out later that he had been twenty-seven years old. For whatever reason – he hesitated, chose not to fire, and I've asked myself many times, "Why?" Then I ask, "But do I carry any regret or guilt for my actions?" And the answer is, "Yes."

I suffered survival guilt for years, not just because of that picture, but because of all the troopers I served with for a whole year. As time went on the fights got more and more serious, and pretty soon I was the only guy of my original unit still alive. I dreaded each day – I was just tired of it, sick of it. Sick of the killing, just sick. I think this particular incident had so much impact on me because it was the first time I'd ever seen an enemy soldier and then it was the first life I ever took. But as time went on, it got easier. I hate to say that, but it did. It became a defense mechanism for me. They were just targets after that. The line between sanity and insanity is real fine, especially in combat. Most of us just tried to block it out, like it wasn't really happening.

Anyway, I carried that picture for years, and my wife would say, "You really should get rid of that, it just upsets you so much." At one point she even asked me about it, so I shared it with her. I knew she was right, but to get rid of the photo, to throw it in the trash or burn it, was just too disrespectful. If I ever got rid of it, it would have to be done with some kind of respect and honor to the people in the picture.

Then, in 1989, I made my first pilgrimage to the Wall. I remember sitting in a hotel room in Fairfax, Virginia, and I knew I was going to leave the picture at the Wall the next morning. I tried to think: if I could say something to that soldier, what would I say? So I wrote a letter. I took the letter and the photograph and put them in a sandwich bag I had with me. The next morning my wife and I went to the Wall, and we left it there and placed a rock on it. Perhaps it was the first time that somebody left something for the enemy at the Wall, but it was catharsis, closure, and I felt good about it.

Richard's story wasn't over. In 1996 a friend walked into his office and said, "Hey, I want to show you this new book. It's called *Offerings at the Wall.* Turn to page 52." When he did, there was the photograph and his letter. It had been collected, like most other memorabilia, by the National Park Service. Once again, Richard was thrown into inner turmoil:

It just – I mean, I'd been at peace with myself, I'd found some closure, and then here it all comes back. It took up two full pages. I just really lost it, was torn up immediately. It felt like it was my fate that no matter what I did, this would keep coming back.

From that day on, I decided I was on a mission to find that family. I knew the odds were against me. I knew it probably wasn't very realistic, but I took the mission on.

Over the next months, he published articles in American and Vietnamese newspapers and contacted ambassadors and veterans' groups in both countries. In time, he received a letter from a man who said he was the son of the soldier he had killed, but when he tried to arrange a personal meeting with the man's family, government agencies on both sides began to stonewall. According to Richard, even U.S. Ambassador Peterson in Vietnam, himself a Vietnam veteran, turned his back on the matter. "They're just totally ignoring the issue. It's taboo; nobody wants to touch it," he lamented. "They're trying to hold reconciliation beyond my grasp."

I wake up at night with cold sweats. You know, my platoon was overrun twice, and I dream about that again and again. And I dream about the little girl. She keeps coming back to haunt me, in my dreams – this little girl in pigtails. If I ever met her, I'd just ask her to forgive me. I don't know what more I could say.

You know, my dad died at forty-eight of cirrhosis of the liver. He was an alcoholic. He was an abusive father, very abusive. With time I was able to forgive him. But try and forgive yourself – that's the hardest part. You can forgive people for transgressions of all types, but just to try to cut yourself some slack...That's the ultimate. It's hard to do. Really hard.

Richard's words made me think of Thich Nhat Hanh, who maintains that such inner suffering can lead to compassion and forgiveness, to a wonderful freedom of the soul.

I talked with a soldier who saw many of his friends die in an ambush, and who then set up an ambush in revenge. Five children stepped into it and died. From that day on, he has suffered terribly. Every time he finds himself in a room with children, he cannot bear it and has to run out of the room. The image of the five dead children follows him day and night. I told him that there are many children dying this very moment all over the world – children who die because they lack just one tablet of medicine. Now, if you can bring that medicine to just one child and save his life, and then you do the same for four other children, you have saved five children.

We must live and work in the present moment. Forty thousand children die every day because they lack food and medicine. So why cling to the past, to the five who are now dead, who have already died? There are more children dying now, and we have the power to change this by touching the present moment.[4]

DAYL WAS IN THE ARMY'S 1st Air Cav in Vietnam, wounded after six months in-country, and spent thirteen months recovering in a Naval hospital in Queens:

When I came back, I kind of retreated. I had several months to think about things in the hospital before I reentered society, and I did what I did to protect myself. I thought it was the easiest way for me to move on – mistakenly so – but, as my family said, I wanted to just bite the bullet: to draw a line in the sand, and then move on. So that's what I did.

I got out of the hospital after a year, but I was still wearing a brace and using crutches. I went to college, and when people would say, "What happened?" I would say, "I was in a horrible car accident." When they asked me if I was a veteran I always said, "No, no. I was 4-F; I never served." That's how I dealt with it.

My son is now twenty-one years old. When he was ten, he had a school project for which he had to interview someone, and when he asked if he could interview me, I thought, "Wow! This is going to be great." So we set up a table and got our stuff ready, and he asked, "You were in Vietnam, right?" and I said, "Yes, I was" – and then he asked me, "What branch of the service were you in?" I was taken aback, but he said, "You know, you've never told me." So there it is. I've been against the war and I have never even talked about it with my own son. I guess that's because it was too difficult to do.

In a sense, I always had trouble moving beyond Vietnam. Vietnam was always with me – it was with me daily. I'd never had a chance, really, to deal with my involvement there as an infantryman in 1970.

I went back for the first time in '93 and then again in '95. That first trip was the result of a project, started in 1991, to collect medical supplies. It was very symbolic, and it enabled me to let go of a lot of things. It also allowed me to deal with my sense of loss – *our* loss – not only of comrades, but also of youth and of innocence. Then I was able to talk to a Vietnamese veteran about *his* sense of loss, and I found that he had moved much further ahead from the war than I had...

The trip helped me to get rid of some of the anger that I had toward my government. And, on a very simplistic level, it relieved a lot of the guilt that I had. I went there as a teenager, you know – nineteen years old, carrying a rifle, calling in mortar fire, inflicting casualties on people, on civilians. And now I'm handing out medical supplies. There's something that's very spiritual about that.

Vietnam's medical situation now varies from area to area, but back in '93 there was a shortage of practically everything. Under the American embargo there was very little ability to import things, and so a lot of the equipment they had was old or broken down. One thing that really struck me was a hospital in Hue. It had electrical power, which is still unusual in Vietnam, but they were still using Coke bottles to hold IV solutions. And they reuse everything, recycle everything – even surgical gloves, which cost next to nothing. They wash them and put them out to dry on little wooden hooks in the operating room. It's amazing...

Then there's the problem of bombs still going off, not only in Vietnam, but also in Laos and Cambodia. During the monsoon season, the heavy rains bring mines to the surface; on our trip this last May, a farmer and two other people were killed just a week before we arrived. In some areas the percentage of farmers who are also amputees is very high. So in a sense the war is continuing.

There are many other personal experiences that stand out in my mind. I met a doctor in Hue who had been with the 33rd North Vietnamese Regiment, and we were the same age. He asked me what unit I was with, and I said the 1st Cav. That really got him going; he knew all about the 1st Cav, and it turned out that we had both fought in Cambodia at the same time, on opposite sides. Just think: in 1970 he was plopping mortars onto my firebase (we even pinpointed one night in particular) – and now we're here, a surgeon and a civil engineer, drinking terrible espresso and eating French bread together in Hue. We were talking about our kids: his son didn't know what he wanted to do, and neither did mine. All he wanted to do was to practice his English. It was beautiful!

Then on my last trip I went to Hanoi, a beautiful, sleepy town of over a million. They have these two big lakes, and every morning around 5:00 a.m. everybody goes down to the lakes to bathe and practice Tai Chi. They're heavily into badminton, and you see these little birdies flying all over the place. I rented a bike and went down there early in the morning, and a family invited me to breakfast. It was very intimate. They were obviously poor, but they served me breakfast and shared what they had. I was very touched, and I tried to give them some money – 5000 Dong, which is like 5 cents – but they wouldn't accept it at all. Those are the things that stand out.

I was always amazed at the love with which I was received by the Vietnamese people. We dropped bombs on them and shot them and laid mines for them. We left children behind us, children for whom we did nothing. And yet I was received lovingly wherever I went. People just invited me into their homes. It was absolutely amazing.

It was very important for me to go back; it's given me the ability to move on beyond Vietnam. Now I can think of Vietnam as a country – it's no longer just a war. I will always be involved in the project, and I plan to go back and spend more time there.

We veterans are afraid to open up the "box" because we're afraid that we might not be able to close the lid. But I've done it, and though the box is still there, I'm now able to close it and move on. Doing something, anything, is an important part of the healing process. It's just essential that it be *done.*

Before becoming a teacher, Jim Murphy was a coordinator for Vietnam Veterans Against the War (VVAW) and was active in the Winter Soldier demonstrations, including a staged takeover of the Statue of Liberty and a gathering at the Lincoln Memorial to protest American military policy in Vietnam. Jim says that if it hadn't been for his early involvement in the antiwar movement, he would have ended up on the street, debilitated and stewing in his own anger, pain, and frustration.

The VVAW movement was the world's largest therapeutic collective. The important thing was that we got together and cried together and talked with one another and worked on each other. I'm sure that the guys who were involved in the early 1970s antiwar movement have better mental health today than the guys who just buried their experiences.

I remember being very angry and very restless. I was drinking a liter of Scotch a day. I had no self-control, and I didn't understand why, but I knew I was angry. At the University of Maryland I ran into a Vietnam vet buddy of mine who'd been a radioman like me. He had been really, really angry over his entire final tour in Vietnam. There was a protest being organized in DC, and we decided to go down and see what these guys were doing. We went to the All Souls Unitarian Church, and their minister was very active in the antiwar movement, involved with other vets, so we just hooked in.

All of a sudden I had a mission. I was surrounded by guys who knew how screwed up Vietnam really was, and we had a goal. The war was wrong, and we were going to end it through this demonstration. This was in 1971, and the demonstration was to take place in April, eight days before the May Day demonstration. So I went back to the University of Maryland and I just started putting up signs. Another guy who had been organizing with Vietnam Vets Against the War was there, and we started a chapter; fifty vets came to our first meeting. All of a sudden I went from being angry and restless and drunk to being focused, with a mission, and surrounded by people who could support me. It was just incredible.

Then Dewey Canyon happened. The famous media picture shows all the guys standing in line, waiting to throw their medals down on the Capitol steps. Guys were throwing back Silver Stars and Gold Stars and Bronze Stars and Purple Hearts and you name it. The steps were covered. I mean, there was a pile of decorations. The federal government didn't have a clue as to who we were or what we were doing, so naturally we were "confused communist Vietnam veterans, drunk and drug-crazed." There were 2000 of us at that demonstration. It was no longer fifty or a hundred; it was 2000 Vietnam veterans in jungle fatigues. The California contingent had M-60 machine guns and bandoleers of ammunition, which was one way to keep the park police away. They didn't mess with us once the California vets got there. And guys just kept coming in all night long, people in jungle fatigues and guys with antiwar sentiments scrawled all over the same boony hats they'd worn in Nam. It was just incredible. It was wonderful.

The park police wouldn't let us camp out on the Mall without a permit. They wanted us over in West Potomac Park, which is as far away from the

Capitol as you can get in Washington, DC. But we camped out on the Mall anyhow, and they didn't dare confront us, because by the next morning we were up to 5000 Vietnam veterans. We had demonstrated to ourselves that we were not alone, and that what we felt about our government and the war in Vietnam was right: the war was wrong; bring our brothers home! That was it. That was the glue.

I got my "fifteen minutes of fame" at the next demonstration, Christmas 1971, when I joined a group of fifteen vets who took over the Statue of Liberty. It was symbolic: this is what you're taking away from the Vietnamese people – their liberty. We made some grandstand statements like, "We won't give up this statue until you bring our brothers home from Vietnam!"

The Park Service had a goon squad, a SWAT team. But they couldn't really storm us, because that would have looked bad – the majority of us were disabled Vietnam veterans.

Our initial plan was to rent boats out on Long Island and make an amphibious landing with five hundred or a thousand vets. But given the amount of undercover agents, provocateurs, and FBI throughout the antiwar movement, we realized that was a little grandiose, and that maybe something small and quiet with good media follow-up would serve the same purpose.

We went in on Christmas Eve. We'd just come up from Valley Forge, where there were about 1000 vets camping out as part of the Winter Soldier demonstration, and we had set up eight or nine simultaneous demonstrations. We painted Honeywell's driveway with blood, and they refused to drive across it. We shut down Caroll MacIntire, the "victory in Vietnam" minister in New Jersey – we just went in and joined the congregation, didn't do anything nasty, but evidently he felt uncomfortable with all of us vets there. We painted the Liberty Bell with blood, and we held a major demonstration at the Lincoln Monument, which was a major bust.

We entered the Statue of Liberty in threes and fours – I'm sure it's much more secure now than it was then – and hid in the arm until after the last boat had left at 5:00 p.m. We had our jungle fatigues on underneath normal clothes, and we all wore our decorations. Anyway, they went after us with a court order, a cease-and-desist order for trespassing, but we were able to hold that up based on the fact that they couldn't get a hold of the secretary of the interior. He was hunting in Alaska, which we actually knew, so they had to go on down through the chain of jurisdiction, and by the time they got down to the New York City level we'd been in the statue for twenty-four hours. By then the Park Service had called a SWAT team in and had surrounded us, but we were barricaded in so well that they couldn't have stormed us without some pretty major property destruction.

The international media was three-deep at all the doors, and there are a lot of doors. The park police couldn't do anything privately; the whole world was watching. So the park police backed off. They didn't want to look bad. So we had control, and we used it. Every one of us was well qualified to speak against the war in Vietnam, and we got a lot of airtime. We were from all over the Northeast: from Boston, from New York, from Philadelphia and from DC, but we made sure that our message was consistent – we didn't go off on any tangents.

In the end, the government dropped it. They didn't charge us with anything. Nothing. They wanted it out of the public view. If there'd been a trial, it would have dragged on forever and we would have been able to make our point over and over and over. We would have loved a trial... We didn't do any damage. We drank all the coffee in the employees' area and ate some of their food, but we left them money for it.

Much later, though, I met Sherman, a guy who was on the desk for NBC News over Christmas 1971, and he told me that they were just getting ready to break the story for the 11:00 news when they got a call from the White House telling them to kill it. Sherman had said, "We can't kill this story," but the White House was adamant and demanded to speak to his supervisor. So the supervisor took the call, and the story was killed. Just like that. By Richard Nixon. And I'm sure that every President does this. It's amazing how much information people never get.

We're still getting revisionist history about Vietnam now, and I think one of the important functions VVAW serves is to make it difficult for the government to lie and change the history of Vietnam. We make sure people don't forget what the real facts were.

Jim is convinced that his activism helped to end the war, but even more important, he feels that it allowed him to connect with other veterans and gave purpose and meaning to a life that was on the brink of ruin. After the 1971 actions he returned to the University of Maryland to study education – believing that if he'd been taught to "think instead of follow," his life might have turned out differently.

Today, he is dean of West Side High School in Manhattan, an alternative school that serves students whose social and economic backgrounds have robbed them of a decent start in life. Jim and his wife Susan, also a teacher, try hard to foster a sense of community at West Side. When I asked them about their work and its connection to the antiwar movement in which Jim played such an active part, he replied:

It's amazing how many of us in the antiwar movement became teachers or other members of the helping professions – special ed teachers, therapists, and social workers. It's been our way to healing.

The kids I work with tend to come from the poorest neighborhoods in the city. So they're the easiest targets for recruiters. I try to help them see what their options are; to help them understand that you don't have to sign up for four years and $40,000, that you don't have to give away four years of your life.

I had a young man, Muslim, who was in the Gulf War. He joined to get money for college, and he was sent to a Muslim country to kill Muslims. He says that they just blew people away. The Iraqis tried to give themselves up, but we just ran them down; there was no time to take prisoners on the front. So we killed them, ran them down, bulldozed them, buried them alive. That's his "be all you can be" story.

The system that educates our kids to accept the military also educates kids to accept all the other things that are wrong with our society. Inherent in having a strong military is racism, and the need for economic separation. If you don't have poor people you can't have privates first class.

I still remember the targets in basic training. The stand-ups we had were "Ruben the Cuban" and "Luke the Gook." Could Vietnam have occurred if people had understood cultural differences? It was a war of cultures and economics.

To create community really brings out the most that you can be. That's where you can "be all that you can be" – in a community that works for public change, for equality. I realize it's a minimal experience, but the important thing is that hope is there.

PAUL PAPPAS SERVED with the U.S. Marines during World War II. He was among the American troops who occupied Nagasaki only weeks after it had been bombed, at the end of August 1945. Paul had been taught to hate in boot camp, and after the war he rebelled inwardly at what had been done to him – and at what he himself had seen and done. He struggled for years to break free from the burdens of his military life, but it wasn't until he faced his own guilt – and stopped pointing fingers at others – that he was able to begin the slow road to full healing.

Mary and I married while I was still in the Marine Corps, and after the war we settled down to start a family. It should have been a happy time. I had a good job, a nice family, two young children. But as I look back, it is clear that I lived, or rather existed, in a constant state of depression. Outwardly, the future looked good. Inwardly, there was no hope, only bitterness and rebellion.

Every person on earth has the desire to be free. But our concepts of freedom can be very diverse. I thought I was free when I refused to allow anyone to tell me what to do, how to do it, or where to do it. This attitude was both blind and arrogant. Now, at seventy-two, I am certain that freedom and inner health depend on just one thing: a clear conscience. And this can be attained by repentance for personal wrongdoing and by forgiveness.

JAY WENK ALSO SERVED in World War II. Like Paul, he has come to recognize his guilt in being a part of what so many Americans still regard as a "good" war. And though he doesn't use the word, his emphasis on tears is very much what Paul's "repentance" is all about.

When I was a little boy of five or six, we would catch butterflies and moths and flies and pull the wings off. I don't look back on that with any pride. I had no sense that I was causing horrible pain to those creatures. Fortunately, I learned somehow, somewhere, that "we don't do that." But a lot of people, I guess, just continue tearing the wings off. Then they end up tearing the arms off people. We graduate from one species to another.

You know, we talk about the horrible Japanese and their internment camps and the horrible Nazis and their slaughtering of people. But there were guys in my platoon and in my infantry company who killed prisoners. And what's the difference between that and My Lai? So they killed women and children as well as men – what's the difference? Taking life is taking life.

I remember our platoon medic. Whenever he had a chance, he would inject enough morphine into wounded German prisoners to kill them. There was another guy, Red, who was told by the lieutenant to take twelve prisoners back to company headquarters for interrogation. It turned out that there was one more than had been requested – the lieutenant had wanted twelve, not thirteen. So he shot one. It didn't "just happen." There were Nazis in our own Army during World War II. And I venture to say they were there any place you cared to look.

What about me? I was a witness to some of that killing, and I never said anything about it, because I thought they would blow me away. Isn't that a war crime? That's a piece of my guilt. I was afraid to say anything then.

We need to get in touch with our tears somehow, some way. And then to continue, somehow, some way. What I'm suggesting is that it can't be done on an intellectual basis. It's got to be done on an emotional basis. I suggest that it starts and continues with tears. Because the pain that I and others have felt about our lives, wartime or not, can lead others to better places. But it takes a lot of time and a lot of work.

As Jay spoke, I had to think of Lee, the Vietnam vet whose story opens this book. Lee says that he can never forgive himself – that he has no energy to love, and often not even enough to live. He contends that he is trying, and acknowledges that his problems are not the only problems in the world, or even the worst – but he feels as though a part of his soul is missing, and he wonders if he'll ever get it back. What would Jay say to a man like him?

I know enough about forgiveness to know that it's exactly like peace, and peace begins at home. Forgiveness begins at home. You've got to forgive yourself, and tears are really the only key I know. If you haven't forgiven yourself yet, that doesn't mean that you can't. But if the tears are still there, let them come. As long as they need to come, let them come.

And if the tears don't come, how do you start them flowing again? Paul Sullivan, a veteran of the Gulf War, says there is no magic formula. But there are times and places in which we can get back in touch with ourselves, with our deepest emotions.

It takes years to deprogram the intensive weeks of military training. It doesn't happen overnight. I would argue strenuously that there is no golden moment…But I could tell you that one milestone was when my daughter was born, and I sat there crying about how beautiful life was, and how on earth could anybody possibly kill anybody.

I could argue that my moment of truth came as I looked at pieces of dead body and blood on my uniform, when I realized that the Pentagon had lied about massive chemical exposures. I could say that it was when a student asked me, "Would you do all this again?" or when someone put me on the spot and asked me to state publicly where I stood on all of this. I could also say it was when I watched slides of the suffering of the Iraqi children and learned that there have been one million Iraqi deaths due to the sanctions. Or was it when I learned that the Gulf War never really ended, that it's still going on? It's all these different things…

We have to recognize, each of us, one at a time, that we have participated in something horrific. And this recognition is the first step toward recovery. Perhaps that makes it sound like I'm an alcoholic, but violence really is an addiction. And as with all addictions, recognizing this is the first step. The next is facing up to what has happened: the deaths and the destruction, the illness and its causes. And then we need to give veterans ways to atone, opportunities that bring about or at least allow for reconciliation. That's not going to be easy, but it's the only way…

CARROLL KING ONLY experienced freedom from emotional torment many years after World War II. He enlisted in the U.S. Air Force two months before his eighteenth birthday in 1942, and by June 1944 he was flying a B-24 bomber in the 15th Air Force Division in southern Italy. His crew continued its bombing raids till the end of the war, striking cities, factories, transport facilities, and enemy troops. Planes all around him were shot down, but Carroll survived, though inwardly he was falling apart.

"I hated the Nazis, and I expressed my hatred from the nose of a B-24, raining down death and destruction on Germany and German-occupied towns and cities," Carroll recalled. Once, his crew even dropped bombs left over from an aborted run onto a manor house in the Austrian countryside, indifferent to the pain and suffering of the innocent people below.

After the war ended in Europe, Carroll was sent back to the States for retraining in a B-29, for the continuing air war in the Pacific. On the ship home, he came face to face with the human wreckage of the war: men without eyes, missing limbs, or disfigured by burns and scars. He was shocked and tried to reach out to them, but was met with bitter silence.

The bombing of Hiroshima and Nagasaki brought a sudden end to the Pacific War, and he came home, went to college, and married. He tried to get on with his life, but the horrors of the war wouldn't leave him. It wasn't until much later that he was struck by the violence in which he had taken

such an active part. His actions and experiences came to him again with blinding clarity, and he recalled the famous painting of the crucifixion by Rembrandt, which depicts the artist himself as one of the executioners. When he first had seen it, the notion seemed ridiculous, but now he understood. The war was in *him*, not in the past, not in the world at large.

He went to a minister and poured out the memories and feelings of guilt from the dark recesses of his heart. Suddenly he felt a peace he never had before. Somehow, in the presence of people he could trust, Carroll experienced forgiveness as a power that cleansed his life.

I had souvenirs, pictures of the bombs, the targets – everybody had them. Pictures of me in uniform. I tore them all up. I trashed everything that had anything to do with the military. I got rid of my uniforms and those medals. In that way, I expressed my determination never again to fight in a war.

I wish I could bring back the people I know I killed, but I can't. You can't undo it; it's just there. And I imagine every murderer feels the same way. I visit men in prison now, and when I hear their stories, I tell them, "You know, in a way I'm just like you, I've done the same things." And I remind them that without God's forgiveness I'm no different than any other murderer. We wish we could bring our victims back to life and ask their forgiveness. But we're not in that position, and so we simply have to accept God's forgiveness.

I trust in the forgiveness I have found – and I really have found peace, by looking into my own heart and confessing the things that burdened me. I've never been tormented by the past again, and I feel that there has been real closure. Of course I still regret the past, and I will always feel the pain of it, but it has lost its power over me.

When you went

to war, you went for the whole nation. The whole nation was responsible for what happened there, not you alone. Your hand was the hand of the whole nation. If you made mistakes, the whole nation made mistakes. If you went to war believing you were doing something important – trying to save a people, fighting evil – it was not your thinking alone; it was the thinking of the whole nation. You were sent there to fight, destroy, kill, and die. You were not the only one responsible. We cannot just shout at you and say, "You did that!" We all did it collectively.[1]

Thich Nhat Hanh

The story of Ron Landsel, like the stories of all veterans of war, is not the story of an isolated individual. It is the story of an entire family, and, ultimately, the story of a nation: a father who died in an army accident before Ron was born, a mother who aged many years when she thought her only son was killed in Southeast Asia, a wife who kept vigil over the long, anguished silence of his "coming home"; a son ridden with the odd mix of guilty curiosity and reverence, wanting to know his father's heart...All were affected, and all suffered the legacy of war.

Unlike the stories of many veterans, however, Ron's is also a story of light and hope. When work on this book began, Ron appointed himself as my "veteran advisor." I had met him and his family some years before, and they'd made me feel very much at home. It wasn't long before I felt as if I'd known this fellow Long-Islander all my life.

But there was something in Ron that never yielded in friendly conversation. His year in Vietnam was untouchable, unmentionable, well guarded. I couldn't understand why, and neither could his family. But there was something that made me want to know, and it was more than just curiosity. Perhaps it was that I'd learned, from my own life, that private pain is the worst pain of all.

Ron was encouraging whenever I talked to him about my ideas for a book, but he never offered to contribute his own story. Only when the project was officially kicked off did the shrapnel begin to surface, piece by piece, as we traded newspaper clippings and swapped addresses. But in the presence of other Vietnam veterans, I noticed that he was beginning to open up. He began to make himself vulnerable by degrees, slowly peeling back the barbed wire that fenced off a whole year of adolescent dread and horror, to find healing in the presence of fellow survivors.

When Ron attended a public forum in which veterans of many wars were invited to share their wisdom and experience with a new and naïve generation, the opportunity for open and honest discussion had a profound effect. Ron's wounds were deep, but their pain could now be shared, lightened, and divided. His relationship to his wife and three children deepened, and he began to reach out to younger veterans with compassion and understanding. And though his journey to full healing may not be over, he feels that he is well on the way: in February 1998, he was able to commemorate the thirtieth anniversary of the Tet Offensive with Vietnamese men and women at the Plum Village Buddhist community in France. For him, it was the link that closed the circle – Vietnam was no longer just a terrible war, but a beautiful country and a loving people. "There is great hunger and need for veterans to find peace and a whole life again," Ron recently said. "And the message I can bring to them, out of my own experience, is one of repentance, acceptance, and forgiveness."

How many families have suffered like the Landsels? Many, perhaps, have suffered even more, but the story that follows ought to give them hope: hope in forgiveness, hope in reconciliation, hope for a new future.

Jason Landsel

I have dreamed of the day you'll come home and finally be my dad.

From a letter left at the Vietnam Veterans Memorial in Washington, DC

I CAN'T RECALL the first time I realized my father was a Vietnam veteran. When I was little we would visit his mom in Bayside, New York, and under the coffee table in her apartment there was always a collection of battered photo albums. One was labeled "Vietnam." It held few photographs, but empty photo holders told of a once-larger collection. The photos left showed little of Vietnam, but rather a young man posturing in military attire. Beneath a few, my grandmother had written things like "Ronnie in San Diego," and in a few other American cities. Other pages had more exotic headings, like "Australia" or "Puerto Rico," but they offered little more for my inquisitive mind.

I had a childish fascination and ignorance about it all – I knew only that my dad had done something unusual, something the other kids' dads hadn't. This was an obvious source of pride and one-upmanship at school: "Well, *my* dad went to Vietnam!"

Maybe Mother had warned us not to, but I can't ever recall asking Dad about Vietnam when I was a kid. There were the little anecdotes he would tell now and then, the light stuff, the stories we came almost to memorize over the years. There was the night a huge rat had crawled over his chest as he slept in a field, the time he was served gin in a teacup and saucer at an Australian bar, the speedboat rides in Sydney Harbor, the nights in a hurricane off the coast of Puerto Rico, and the fact that he almost died of malaria. This is what I knew.

It was only when I was older that I began to wonder about his experience in a serious manner. He still never said a thing, but my mother would spill a few details now and then, mostly from the time when he first came home: the silence; the nightmares; the late-night promise, made to God on his knees, that none of his sons would ever fight in a war...I was deeply impressed, and I wanted to know more, but I knew I would never, ever, dare to ask.

Over the years I schemed to get his story, feeling that it was important to document his life – if for nothing but for his family. I wrote him lists of questions and even hinted at taped "interviews," but nothing happened. And I felt guilty planning such projects, fearful that he might "lose it" – that my questioning might bring back something from the past to trigger a very negative reaction. I even discussed my plans with my minister, hoping to get his support, but was told that "it wouldn't be helpful to bring up the past."

Then, last summer, after we had worked together on an event presenting the truths of war to a group of young people, he spoke to me about Nam for the first time. We had watched *Platoon,* and it had been a long evening. We went outside to get some fresh air, and I hesitantly asked him whether or not the film was authentic. His answer was to tell me of his own experiences – general information at first, but it was something: the sound of the bullets, the food, the missions. I probed on carefully, not wanting to stop him, aware that I was being entrusted with something that was very deep, very personal. Never before had I felt that close to my father. Then our walk was over, but our relationship had been forever changed.

The details of his trip home stand out in particular. The ride from jungle combat to suburban Long Island took less than forty-eight hours, and after Vietnam, life in New York was more than he could bear. He remembers sitting in his hometown mall, watching the hurried and self-absorbed shoppers, wishing he could just "blow them all away." I told my mother about our talk late that night, and she agreed that it was a very special moment.

There were other times that summer when he told more – unexpectedly, to be sure, and with obvious difficulty. But I saw in his stories a man changed, and in spite of the strong emotions and unusual profanity, I gained new trust and new respect. I had been given a glimpse of untold pain.

That summer I read all I could about Vietnam. Reading the accounts of other veterans, I realized how lucky our family had been. Dad had started out no differently than the men I was reading about, but by some twist of fate we had been spared the ongoing misery that is the lot of so many veterans and their families. And when I considered the statistics of divorce, abuse, suicide, and alcoholism, I knew it was nothing other than a merciful God that had steered us to where we were now.

Without God's help, and the help of a committed and supportive community, there's no way my father could have overcome the violence of the war. And I pray that many others might be granted the same mercy.

Pat Landsel

Women did not go to Vietnam in great numbers, but great numbers have been scarred by the war. One by one, without the awareness of danger that strikes those who put on uniforms, we met and fell in love with men whose combat experience would change our lives... The men of our generation were called to Vietnam: some went off to war, some of them came home. We thanked God it was over. And wearing frilly-nightgown smiles, we slept snuggled next to walking time bombs.[2]

Wife of a Vietnam veteran

Ron and I met when he was already in the Marine Corps. In those days, Marines were admired. To be in the military was normal and expected of a young man growing up.

My father was wounded in the Second World War and received a Purple Heart. He came home to a very young wife, and for the rest of his short, hard life, he suffered a serious drinking problem. He died very suddenly at the age of thirty-two, leaving behind three young children. I was seven.

My mother married again. My second dad had served in the Navy; later, my brother Joe served in Korea. I grew up on movies that gloried in the fight for Democracy – the fight against the evil communists. And as a Catholic, I learned early what a privilege it was to fight for one's country.

Ron stopped at our house on his way to Vietnam, and over a Scotch-and-water, he asked me to marry him when he came home. I said yes, and together with his mother and a friend we headed for the airport. Neither of us saw the incredible step that he took at that moment.

Then he was gone, and I wrote him faithfully, sitting in his car which he had left in our driveway. I sent pictures and tapes, and went often to the parish church to light candles for his safety. My walls were covered with his letters and pictures, but I never asked once what the war was all about. And I never questioned my country, or the President. Even after Kent State,* after the cries to "bring our boys home," I didn't see what was happening.

Ron served a year in Nam, and I went with my dad and his mom to the airport. All of a sudden, there he was, holding me, kissing me, but not saying very much at all. Over the next weeks we spent as much time together as we could, but he slept and slept and slept. I didn't know what to say – I didn't know how to help or support him. I never asked him about Vietnam, either, and he never said anything. He was tough, he was the all-American Marine, and he could handle it.

I had made a large poster from a picture he had sent from Vietnam. He was sitting on sandbags, holding a machine gun – a really tough picture. But when he saw it, he made me take it down right away. I did, but I never asked him why it bothered him so much.

We saved up and had a dream wedding, but we were nothing special. We were selfish young adults, and there was nothing in particular that held us together. Then he came home drunk one night – he had damaged the

*On May 4, 1970, the National Guard was called out to quell a student rally at Kent State University, Ohio. In the ensuing confrontation, four young people were shot dead. The incident provoked nationwide protest and controversy.

car a little – and I remember sitting on the stairs and looking at him as he said, over and over, "I'm crying, I'm crying, and I haven't been able to cry since the war." I felt real sorrow for him, but again I didn't know what to do or say. I wish I could relive that moment now.

Another time he was yelling, and he said, "If we have sons, I'll never let them go to war." I was scared when he said that, because it might mean we would have to leave the country. But he still said nothing about the war.

Both our lives had been wounded from early on, and neither of us knew what it would take to make it through, to keep our vows to each other. But then we experienced a life-changing conversion; we opened our hearts to Jesus' words, and we became radically serious about following him in our everyday lives.

At that time, our son Jason was born, and things started to change fast. We renewed our marriage vows and dedicated our lives to God. But the pain was still there, and Ron said nothing about the war.

After twelve years in a Catholic community attempt on Long Island, we joined the Bruderhof movement[3] and we were confronted right away with the nonviolence of Jesus' teaching. It struck us immediately – "Oh my God, of course!" Here was a life of love, lived by people who were seriously committed to what Jesus had demanded for the whole world.

Still no sharing about the war...

Even within a supportive community, our marriage and our lives as individuals were in need of continuous renewal and recommitment. We continued to hurt each other, our three children, and those around us. We had tough years, but because of our clear commitment to God, and our willingness to work out any problem between us, however painful, we made it through.

But Ron *still* said nothing about the war, and it wasn't until he wrote his account for this book that the whole story came out. He labored over it for months, and it took much prompting and many prayers. But there was a purpose to it all, and I pray now that many more veterans can be granted the healing and new life he has found. His story is about the horror of war, but it also points to a God who cares.

Ron Landsel

"Back Azimuth"

War's unreality
Is horrible
Inseparable
Languishing
Silently
In faded rooms
On the other side of a latch
Just outside a window
Imminent
It will lift
One of many mask-faces
In terror's hoary remembrance
Knowing your deepest emotions
Your limits
Never needing to speak
You sigh
Ron Landsel

THE FIRST DATE ON my tombstone was recorded on August 16, 1946, just a few blocks from the star shows of Hayden's Planetarium, uptown, Big Apple. Shadows of World War II nightmares remained a secret for millions, but not to Mildred H. Landsel or to the tens of millions who suffered in the world's new killing fields.

Little-known brush fires in a remote area called Indochina were already flickering on the newsreels, fed to us growing boys at the 25-cent Saturday matinee between the cartoons and heroes that mustered us out into games of "playing guns." We divvied up into good guys and bad guys in the empty lots of working-class Queens. Fellow players considered my mock deaths a great contribution to the game. My cold-war MOS (military occupational specialty) in Catholic grade school was "window-shade monitor" during air raid drills. As the widow's son went to hide under his desk after "protecting us from the bomb-shattered glass," Sister Mary Ronald, my heroine and the consoler of my insecurities, knelt and prayed with us for the conversion of Russia. I loved her kindness and hoped her prayers, and the benign Dwight David Eisenhower, would make us safe from "the enemy."

After school, Sir Baden-Powell's followers, Scout merit-badge courses in hand, prepared us for the field disciplines of tracking, communications, first aid, compass, and topographical map skills. Our weapons: the war-surplus store's sheath knife, entrenching tool, compass, semaphore flags, combat pack – and ignorance, the lethal seed of war.

Except for pointing out a neighborhood man as a World War II flying ace, and rare comments like "it was good to have a general for President," Mom never talked about the war or politics. Her love, Earl, "died in an Army accident just before you were born." (I learned later that over 10,000 deaths in Vietnam were due to accidents…) A Cornell-graduate and a lithographer, he and Mom, a physician's adopted daughter, hailed from the lake region of Syracuse and Buffalo.

Mom never remarried, and we were a lonely pair, comforted for the next nineteen struggling years by precious little of Dad's artwork or his life story. We talked little about our pain, and never about his death. Nor did we talk, years later, about the terrible day – the shock and then the relief – when two Marines in dress-blue uniform rang the doorbell to Mom's second-floor Queens apartment. Did she somehow already "know"? Did she think, at that moment, about my father and another war? The news was not as terrible as the setting: I'd been nearly a month in the hospital in Da Nang, but "the nice young fellows" assured her I was recovering well.

I wonder sometimes if I didn't nearly die in that hospital, if some power didn't reach down and push me back into life just as they were speaking to my mother. Vietnam presented me with several experiences of what I now know as eternity.

We were working-clàss poor, with nothing much beyond the basic necessities to speak of. I was never the best-dressed kid, and I vividly recall the long discussion that Mom had with Father Terence Sharkey, the retired businessman-turned-priest, about first-grade "tuition" – whatever that was. She told me nothing, but in September I found myself in the first grade at Sacred Heart School in Bayside, New York.

In the eighth grade, I went to work as delivery boy for an Italian butcher and grocer. I worked every day after school, and eight hours on Saturdays. I earned eight dollars a day, lunch on Saturday, and tips to boost the ninety dollars Mom made as a secretary at the air conditioner factory. As I sorted smelly beer and soda bottles in the dingy store basement on those long Saturdays, I quietly perfected my broken English-Italian accent, considered the life of the martyrs, of runaways, and whether the hottest thing in town, old man Doyle's basement fallout shelter, was motivated by his World War II experiences. He had the whole deal: lead-lined walls and ceiling, canned water, and shelves full of food. His son secreted us neighborhood boys in for a peek; we were buddies and "on the list."

The neighbors offered painting, snow-shoveling, and mowing, but we were still pretty poor. I remember stealing a crisp five-dollar bill from a friend, and needing lots of help to face it. Just around Easter, a neighbor hanged himself one Saturday night in the back of his hardware store. Man, that rocked my life. He'd had a good job and was always around. We'd always admired his family; his son was just a little younger than me. But he stopped coming to our school, and I felt terrible. I didn't want him to be without a dad. Mom told me that he had fought in the war…

I was crushed when my first love left me after I told her I wanted to be a priest. She was also upset because I never kissed her. My feelings of inferiority had not brought me to such achievement; carrying her schoolbooks and sharing dreams on the long walks home from school were wonder enough for this latch-key kid.

A basketball scholarship came along by the time I was ready to enter Holy Cross High School in 1960. My Catholic Youth coach, Mr. Miller, was a great encouragement. But high school moved too fast; and although I loved art and music, the pressures of basketball, football, beer parties, girls, and fraternity life came rushing at me in those short years of getting an education. My independence cost me my virginity by the end of the tenth grade.

One day, as I stood in front of our favorite after-school luncheonette, a young black guy was shot right across the street by a chasing cop. Other cops kicked and beat him as they waited for an ambulance; I wanted to shout out and stop them because he was obviously hurt. But I was already learning the moral "freeze," and though I was stunned and angered by this violence, I did nothing.

I was sixteen when a very tough neighborhood kid lost his legs in some place called Vietnam. It was all over town: "A booby trap – blew his fucking legs off." I was sickened and wondered why we never saw Tom again. And though it seemed that the prayers of Sister Ronald and Cardinal Spellman were keeping the Russians away, what about the commies in Vietnam?

Mom was always there, quiet but supportive. She pointed me toward the church, though she rarely went herself, and there was the copy of Kahlil Gibran's *The Prophet* that I was supposed to read when I got older. She gave herself tirelessly, coming home from work, shopping, cleaning, making supper, and sleeping on the couches of our one-bedroom apartments – all her life. And only rarely did her calm, which hid a deep well of pain, erupt in bitter outbursts – usually when someone had hurt her.

Only days out of the combat zone, I came home to my second-floor apartment in Queens and lay down for three solid weeks of sickly sleep. I also conversed with the ghosts that had accompanied me home, which try as I might would not relax their grip on my soul.

I was looking down at my shoes when Mom, Pat, and her dad came to Kennedy Airport to pick me up. That was weird; I felt like a movie actor who'd forgotten his lines. I just stood there grinning, mumbling innocuous "hi's" and "how are ya's," hugging, hoping everything was going to be okay. I was freezing, and Pat's dad covered my tropical uniform with his coat. Then we walked to the car for a very difficult hour-and-a-half ride home. What do you say after the clichés and the standard explanations? I lied, because I no longer believed them to be true. I was supposed to be a hero, but I felt like I was riding in an open coffin. My family stared at the corpse, occasionally stuttering, but afraid to break the silence. The DJ on the radio sounded like an empty suit. Every sense was assaulted. Sounds, sights, and smells. It was all wrong. This was not what I had fought for.

By that time, though, I was morally compromised by the values I lived for. Selfish interests, the easy life, women, drinking – all my posturing in Catholic-American loyalty denied the response of my own heart, which knew it was all wrong. And it seemed that the Marine recruiter who'd gotten me into this hadn't listened to his heart, either. He never told me that I could easily and legally have avoided the conflagration as a sole surviving son.

Pat was my constant in this lost world of my peers. She was the "back azimuth" that I could follow to safe haven. She didn't ask the stupid questions: "Did you kill anybody?" "What is war like?" She knew I would never be the same again, that there was no point in telling me that it was okay.

It wasn't. Something in me was gone. Dead, but not yet buried, and wandering, like the *mat tich,* the hundred thousand wandering souls of the missing Vietnamese war dead.

I was worried. I couldn't concentrate very well anymore, and the things I had been through in Nam were breaking into glimpses like the pieces of a broken mirror. The days and months there no longer fit together. I was easily exhausted. Malaria had taken my stamina, and still keeps it. I ignited in rages over things insignificant, and I kept everything inside. I desperately wanted the war to be over, but it was everywhere, and still in my own sickened spirit.

The Vietnamese find peace through a rootedness to their ancestors. They revere their forefathers and keep little altars of remembrance for deceased loved ones in their homes. But there were no altars of reverence in our society – not for the dead, and certainly not for returning veterans. We were ignored in the job market and on the street. We were spat upon, beaten, and denied our promised benefits. There was no incense for us. We symbolized an America shocked into awareness of our lifestyle's horrible effect on the rest of the world. We fought for its corrupt system and were hated for it.

"Ask not what your country can do for you…" That call had been an appeal to survival, to community, to a brave new future. But the America I fought for in 1969 was more interested in watching the N.Y. Jets-Baltimore Colts Super Bowl, and standing in long lines at malls returning unwanted Christmas presents. And worried about the upholstery in their cars.

I wanted to draw from under my cloak the bloody, heavy pieces that my soul had become. I wanted to hold out the dead, the POWs and the captured VCs – to stand with them in front of the 21-inch color TV sets, sit with them in the shiny new cars, hold them up in the department-store windows for everyone to see: stumbling, tagged, and blindfolded, as their "liberators" led them to the punishing winds of the huge CH-53 choppers that would take them away to refugee camps in the South…

When the bartender asked me what time the game started and who was going to win, I wanted to scream out: "When I was on duty, not one capture was shot or raped!" And it was true – I had refused orders to call in fire on a group of women. I dumbed the barman's stupid questions, stirring scotch and ice with my trigger finger, thinking that a tour in the Nam might do this guy some good.

I came home from a war very few Americans cared about or understood. It was the TV war: "Oh yeah, I saw that Tet thing on TV – nasty business." It was, but they hadn't drawn its pain down into their lungs, holding the hit as long as they could. Their noses, throats, and eyes hadn't been seared by the heat of terror, grief, anxiety, and rage. They hadn't heard the sucking sighs of the dying, hadn't seen the pink bubbles and the flapping gauze. They had just turned the channel while I rock-and-rolled* ten magazines into an early morning ambush that took out a quiet new kid from Wisconsin. He was hit only inches from me, and his life and hope and love gushed thick from his mouth and nose as we followed each other's eyes all the way down to the jungle floor. He never said a word, and neither did I. Why wasn't it me? In a way, it was.

I'd already seen a KIA (killed in action) in boot camp. We diagnosed his death as terminal fear, but rumor had it he died of pneumonia. He'd been sick for a while but was terrified of the harassment you could get for going to sick bay. It all came back now: "Kill a commie for Christ!" "I can't hear you ladies! Are you my ladies?" "Sir, yes, Sir!" "Who you gonna kill, ladies?" "Sir, commies, Sir!" "Who for, ladies?" "Sir, for Christ, Sir." "Bullshit – I don't believe you – bends and thrusts forever. Reaaady – begin!"

I came off battalion OP (observation point) too sick to go out on patrol. I'd had a wicked fever and no strength for over a week. Our sick bay couldn't come up with anything, but I couldn't walk or eat anymore. I just shook and burned. Then the high-pitched whine of rocket-propelled

* Full automatic fire from an M-16: about 750–900 rounds per minute.

grenades came into base camp, and I had to be dragged to a bunker. Next morning I was medevaced to a unit near Freedom Hill. I had a temperature of 106.8. Malaria. I was going to die lying there in the ER. A dustoff brought in three grunts and a child. The child was like a skinned animal; he had been dragged behind a Six-By, and died on his litter. One of the grunts had been hit in the abdomen and was screaming; they were catheterizing him when his heart stopped. I remember his body lifting up off the cart, again and again, as they defibrillated him. He died, too. How did the medics keep this up all the time? Then I was strapped across the back of a jeep and taken to China Beach.

At the Da Nang Naval Hospital, I was thrown into crushed ice baths, blasted with floor fans, and pumped with medication. The one rocket attack I remember during my stay wasn't as scary as the ice treatments and the allergic reactions I had from the medications. A doctor told me I was going home, but the big boys had other ideas. I had recently been promoted to sergeant, and they reassigned me as 1st recon communications chief for Delta Company.

After boot camp, I had trained in combat and radio-telegraph operations on the East and West Coasts. Then, at the kind invitation of General William Westmoreland, who asked for 206,000 more troops in 1968, I had gotten orders for Vietnam. Martin Luther King and Robert Kennedy were both slain that year. History was screaming at us, but nobody listened. I still remember the flight to California when reality first hit. I sat there and began to cry, silently. No one said anything, but the well-dressed Asian businessman next to me got up from his seat and moved.

But I was in denial right up to the moment I stepped out of the air-conditioned commercial airliner that landed in Da Nang. No one from training was with me, and my own countrymen didn't know me, nor I them. Like the Vietnamese workers, they were callous and careworn – we had little in common beyond our births on the same huge continent.

Their slang, indifference, and even threatened violence were a painful destabilization. And the terrific heat was like a beast drawing me into its den. I was convinced that I had come here to die, and assured of my premonitions when I got assigned to my new unit.

The touted life-expectancy of a Marine radio operator in a firefight was about eleven seconds. Entire reconnaissance teams were given half that. There were always three dead men in a recon team: the point man, the patrol leader, and the man with the antenna. But like the men around me, I kept my fears to myself. No one voluntarily exposed his fears. From the grunt to the North Vietnamese regular, from the Vietcong sapper to the boys in Hanoi, all the way to the State Department – we were very courageous warriors. There was plenty to bitch about, but we hung tough in our fear. And so the wars of our minds lasted longer, and were as destructive as the exploding shells and the bullets.

In-country recon meant training on a battalion OP, in a so-called free-fire zone southwest of Da Nang. Our position commanded a full view of the river and valley below, and we ran constant patrols off the hill as well – the Vietcong and NVA were constantly moving into the area. I got plenty of experience calling in air and artillery strikes, including *Puff the Magic Dragon,* a slow-flying transport plane equipped with Vulcan machine guns that could saturate every square inch of a football field in minutes. We would watch the lights of its tracers in dumb amazement as it flew up and down the valley for twenty minutes at a time, spitting red fire...

It was Christmas 1968, and it was the pits. Bare, dusty, blown-up red soil, stony rubble, trash waving in the concertina wire fences. This place was hell. Brown rats crawled over our chests as we feigned a night's sleep in the bunkers. They reminded me that there were brown men all around us who could easily do the same. Our platoon was not overrun, but we were randomly mortared – sometimes heavily. Our bunkers were no match for their shells. But their ammunition was precious, and unless they locked in on us quickly, they didn't stick around to wait for the artillery support we would call in as retaliation. All the same, the silence between explosions was infuriating. They seemed to be right on top of us, and then – nothing. Just our heartbeats. It was torture.

One side of the OP was often subject to sniper fire. I was standing in a trench one very hot afternoon with two other guys when rifle shots came in. The first round hit a hand's width in front of my chest. Years later I could see that nightmare round, going through my head, spinning me around as I took two dead steps away…

2:00 a.m.: a repeated radio message from an OP across the valley saved us from being overrun. They didn't know our call sign but still managed to get our attention. They had seen beaucoup lights ascending the steepest escarpment of our hill, but to call in artillery was useless; these guys were doing some serious rock climbing. We set out extra claymores and impatiently tore open cases of hand grenades. As we lobbed them over the edge, the lights fell and went out. We called in Willie Peter (white phosphorus) at the base of the hill and stayed on 100 percent alert the rest of the night – our silence spoke the memories of being overrun before. I radioed in a schedule of the sickly-white illumination rounds until dawn. We didn't speculate too much about what could have happened, but we knew they had to be sappers, climbing a ledge like that. If it weren't for the radio call and the clear night…The little parachutes of spent illumination rounds hung lifeless in the dead trees, like symbolic ghosts.

Reconners hated OP. We were stealth, silence, jungle – the cat. We never used trails, streams, or low ground. But patrols off this hill offered little safe cover. Ambushes and mines were not infrequent encounters on the worn trails up and down the slopes. My first patrol was a total screwup. We got down to the valley floor beautifully. I was quiet and alert, the only new man; it felt right. But then we had movement in the thickly vined brush ahead. The team leader signaled, and we nudged our M-16s off safety. My heart was pounding as we crept forward. BLAM! Everyone was on their bellies in a circle, in a split second. I'd had an accidental discharge, and "Country," the team leader, was ready to blow me away. His rifle muzzle shook an inch away from my nose as my round echoed along the valley floor. I could never do that again to him and live.

Afterward, we did several successful patrols together. He was good, and I learned a lot from him. We became buddies but partied separately: he liked country music and cold beer; I did Hendrix and pot. It was hard to say goodbye when Country rotated. I wrote him once, but I never got in touch with him stateside. I was reluctant to reach any of the guys, for that matter. I didn't want to go through it all again.

Recon patrols were dangerous, frustrating work. Insertions and extractions by helicopter were occasions for ambush by the Vietcong, and I lived through three of them. One correctly placed bullet could bring a helicopter down, and even if you lived through the insert and made it to high ground, each hour was still pregnant with death. There were mines, and even bad weather could get you killed. One of our force teams was totally wasted – wiped out – by lightning. Heavy fog and rains prolonged our patrols beyond coordinated extraction, and more than once we were forced to move on to an alternate LZ (landing zone). On occasion, we had to use a precious claymore to blow an area open ourselves. This was the nastiest way out, risking an invitation to the bad guys to see us off.

We only carried M-16 rifles and one M-79. No mortar, and no M-60 machine gun. We did not wear the normal flak jacket and helmet, and there were no ponchos or blanket-liners to soften the cold night-rains. We took no extra food and went hungry if socked in; once, we humped three days on stream water. We slept on razorback slopes, up against rocks or trees. Our stealth and our face paint were our only security, and as night surrounded our tiny, eight- or nine-man team, we would eat our cold rations or tinned fruit, assign watches, and commit our grid position to memory – just in case we needed to call in artillery around us. It was a farewell ritual. Only "Situation Report," the voiceless keying of the radio handset on the quarter hour, told the relay on Monkey Mountain that we were still a part of creation.

Exhaustion numbed the day's fears but magnified the images and sounds of night. One night, on watch near the border of Laos, I woke the team to hear the horrible sound of hundreds of troops in the underbrush. We listened, frozen, as we double-checked our memories for the grid coordinates. They came closer and then stopped. Now they were walking around in circles. Then they stopped. Were they going to spend the night this close? We dared not make a move. They were too close, too many, to do anything but make deals with God. Then, after what seemed like an hour, we heard them again. Elephants! A herd of freaking pachyderms!

It was late in the afternoon of June 21, 1969, when we got a red alert that a recon team was compromised and being chased and fired on by a large enemy force. We pulled an emergency reaction team together from Delta Company; they were ready to go within minutes. Second Lieutenant Bill Schenck of New Jersey and Lance-Corporal Joe Bosco of Providence, Rhode Island, headed up the patrol.

Two hours later, we received the crushing news that our Delta team, along with their CH-46 crew, had been ambushed on insertion. All were killed. A third team had been unable to insert in the vicinity because of heavy ground fire, and our chopper was in flames. Secondary explosions from the craft meant that the mines and grenades they carried on their backs and belts were being set off by the inferno. It was getting dark and S2 lost radio contact with the first team. It was a long, painful night. I was just back from Da Nang, recovering from malaria, and I nearly made this trip. It would have been my fourth and last time in ambush.

At home, a Gallup poll revealed that 55 percent of Americans opposed the war. On October 15, 1969, more than 10 million Americans across the country demonstrated their opposition to the war. In November, 200,000 people marched in San Francisco, and half a million more converged on Washington's Mall...

We have a .50-caliber machine gun on our OP. It's a horrible thing to use against men: it takes two hands to fire, and it rips off chunks of flesh and bone. Its recoil travels through my upper body and sets me to rocking with every round fired. If they wear black pajamas, I open fire. Conical straw hats fly as they dive for cover. They're pinned down now, behind a huge fallen tree near the base of our hill. This is madness. Free-fire zone? What am I doing? I'm going insane.

We are tunneling down through the canopy into view of the early-morning jungle floor. The *Sea Knight* drops her dragonfly tail only five feet from the wind-flattened grass. We jump as one – like young from the belly of a dragonfly. Our canvas gear dances into a hasty perimeter. Seconds pass. Prone. Silent terror. Then the CH-46 lifts her tail, and as she turns, the brittle snapping of rounds sounds overhead. She opens with withering M-60 fire – 550 high-velocity rounds per minute – and we unleash our M-16s into the same direction. The chopper lowers once again, firing into the ambush. We crawl backwards, shouting our names, dragging, pulling one another back on board. A thud, and our chopper is hit. But we are covered by a swarm of Huey gunships, strafing and destroying their position, and as we lift off, I radio in: Two Oranges, Zero Apples."*

Actually, we all died a little more that day.

Nearly five thousand helicopters went down during the war, and all but a few of them had crews on board. I have reconciled with the war. I respect and love the Vietnamese people. But there are certain things I still cannot stand, and one of them is the sound, sight, or smell of a military helicopter.

---

* Two wounded; none dead.

Even before I ever saw action in Vietnam, I was assigned to a burial detail in North Carolina. The funeral ceremony for James was held in his poor-as-dirt Pentecostal church. There we were, five crackers and one black, sitting right in front of the pulpit, where hope was spoken to the inhabitants of little more than tin and board homes: "Oh, yes, brothers and sisters! Jamie's life was poured out for freedom! Hallelujah!" shouted the black pastor. The worship and song was highly emotional. It had to be. The open casket held no warrior – he was such a small, young boy. Church attendants fanned those passed out in heat or ecstasy – the pain of poverty, repression, and now the pain of death. It was more than I had realized up till now.

We took James's casket out into a field of his family's small farm. The heat of the day was broken by a strong, steady rain, and I remember the weight of the rain-soaked flag as I extended it to James's mother in proper ceremony. Her eyes did not meet mine; her pain was too tangible. Was she thinking of her boy working in this very field just last spring? And then we realized that a terrible mistake had occurred: as if to salt the wound, the casket did not fit into the marble-lined grave site. Even today I ask myself, would it have helped if I had gone back on my own and dug the grave larger? Would they have believed that the government really cared?

There are other stories: the rocket attacks, the wretched night-ambush missions. But the hardest ones, the daydreams that are still a part of me – those are my real war stories, because they are obscene and evil.

The battalion sergeant and I are out on the town. Top knows the older woman – the young girl's mother? Their sunlit room rings with giggling and laughter, as I and the child-woman (she couldn't have been more than sixteen) were thrown together, trembling with the horrible business of war...Hate for myself, and the incessant memory of this young girl, left the house with me. She was the victim I took as a hostage to my loneliness and selfish lust.

Many tales of violent fighting, bloodshed, and terror could still be told. But they pale in comparison to that quiet, sunny Sunday in residential Da Nang, that brief, wordless moment when the screams of two souls echoed in eternity.

In a large part, the Vietnamese have forgiven us. They love their roots, their people, and their country too much to hold on to hatred. They have borne many more wounds than we, but their suffering has brought them understanding. And we can learn from them, because we understand, in part, the same suffering. We can rebuild our lives in the same way that they fill the bomb craters in their villages: one basket of soil at a time. They don't dig much from outside the crater, because they have learned that the soil is deeply compacted from the force of the explosion. So they loosen it, and basket by basket they level the earth from within. We can heal, too, by gently loosening the clay of our own hearts. We can fill the craters of our souls, basket by basket.

When I returned from Vietnam, I was angry and confused. I had no understanding of how to become free of my burden, so I turned to the alcohol and marijuana that I had learned to depend on in the military. I had no peace, and I sowed the seeds for more war. I became more angry, more guilty, and more confused. One night, in a drunken argument over the war, I found my hands clenched around the throat of a very good friend. My burdens were destroying friendships, our marriage, and slowly killing me.

Now I am at peace. I can say this only after a long journey into the awareness of Jesus and his living truth. I believe that my fellow survivors can learn to know Jesus just as fully – can follow after him in their own lives and identify themselves with his own historical concern for peace and justice. But I'd say that you have to live for something else, something bigger than yourself, bigger than your own problems. There have to be deeds motivated by the spirit of his love. Giving my life to something more meaningful, to working for peace and justice, for harmony between all people, is the only way I can deal with my past.

War is not the beginning of evil in our society; it's the result of evil. And I've been a part of it – I have to look at my own responsibility. Fortunately, I've been given the wisdom of repentance, seeing my complicity in the pain of war. And I've learned to trust in a power that's available to each and every one of us. It's the power of forgiveness that Jesus speaks of in the Lord's Prayer: "Forgive us our trespasses, as we forgive those who trespass against us."

Endnotes

Chapter 1

[1] For a brief description of the origins and role of the U.S. Department of Veterans Affairs (formerly the Veterans Administration) see chapter 4, note 1.

Chapter 2

[1] Dalton Trumbo, *Johnny Got His Gun* (New York: Bantam Books, 1988), "Introduction."

[2] *Chronicle of America,* Jacques Legrand, publisher; Clifton Daniel, editorial director (Chronicle Publications, 1989), 584.

[3] *Ibid.,* 587.

[4] Howard Zinn, *A People's History of the United States* (New York: The New Press, 1997), 265.

[5] *Ibid.,* 263.

[6] John Steele Gordon, "What We Lost in the Great War," *American Heritage,* July/August 1992.

[7] Michael Bilton and Kevin Sim, *Four Hours in My Lai* (New York: Penguin, 1992), 8–9.

[8] *Chronicle of America,* 603.

[9] *Ibid.,* 589.

[10] Howard Zinn, 267.

[11] *Chronicle of America,* 603.

[12] *Ibid.,* 599.

[13] James Trager, ed., *The People's Chronology: A Year-by-Year Record of Human Events from Prehistory to the Present* (New York: Holt, Rinehart & Winston, 1979), 758.

[14] Dalton Trumbo, 228–232, 240–242.

[15] L. A. Dietz, "Veterans Day Service at Fusa," Nov. 11, 1990.

[16] *Battle for the Falklands* (video), Independent Television News Ltd. and Granada Television International Ltd., 1982.

[17] Ken Lukowiak, *A Soldier's Song: True Stories from the Falklands* (London: Orion, 1997), 179–180.

[18] Wendell Berry, *Sex, Economy, Freedom and Community* (New York: Pantheon, 1993), 70–71.

Chapter 3

[1] Joseph C. Conrad, "Environmental Considerations in Army Operational Doctrine," White Paper, CR-9421, January 1995.

[2] Tim O'Brien, *The Things They Carried* (New York: Penguin Books, 1990), 21.

[3] Anna Simons, Review of *Making the Corps* by Thomas E. Ricks, *The Christian Science Monitor,* December 15, 1997.

[4] Bob Muller, "A Veteran Speaks – Against the War," *The Fight for the Right to Know the Truth: A Study of the War in S.E. Asia Through the Pentagon Papers and Other Sources* (monograph) (New York: The Student Assembly of Columbia University, 1971).

[5] Iris Chang, "Exposing the Rape of Nanking," Book excerpt in *Newsweek,* December 1, 1997.

Chapter 4

[1] "To care for him who shall have borne the battle, and for his widow and his orphan…" These words from President Abraham Lincoln form the motto of the Department of Veterans Affairs (VA), formerly the Veterans Administration.

The VA, which traces its roots back to the state-run homes and medical facilities for veterans of the 1800s, was established by Congress in 1930, and in the years since then has been designated the responsibility for virtually every matter relating to veterans – from pensions and benefits to medical care to tending the Arlington National Cemetery.

According to the VA's official website (www.va.gov), its health care facilities "provide a broad spectrum of medical, surgical, and rehabilitative care" that includes "171 medical centers; more than 350 outpatient, community, and outreach clinics; 126 nursing home care units; and 35 domiciliaries."

In 1989, the Veterans Administration became the Department of Veterans Affairs, established as a Cabinet-level position. President Bush hailed the creation of the new Department saying, "There is only one place for the veterans of America, in the Cabinet Room, at the table with the President of the United States of America."

2 *Trauma and the Vietnam War Generation: Report of Findings from the National Vietnam Veterans Readjustment Study* (New York: Brunner/Mazel, 1990).

3 "Vietnam Combat Stress: Linked to Many Diseases 20 Years Later," *Advance for Occupational Therapists,* November 24, 1997.

4 *Burning Conscience: The case of the Hiroshima pilot, Claude Eatherly, told in his letters to Günther Anders* (New York: Monthly Review Press, 1962), 80.

5 Martin Gilbert, *The Second World War: A Complete History* (New York: Henry Holt, 1989), 712.

6 *U.S. News and World Report,* July 31, 1995.

7 *Ibid.*

8 John Hersey, *Hiroshima* (New York: Alfred A. Knopf, 1993), 68.

9 *Chronicle of the Second World War,* Jacques Legrand, coordinator; Derrik Mercer, editor, (London: Chronicle Communications Ltd., 1990), 652.

10 Ronnie Dugger, *Dark Star: Hiroshima Reconsidered in the Life of Claude Eatherly of Lincoln Park, Texas* (Cleveland: The World Publishing Company, 1967), 115.

11 *Burning Conscience,* ix.

12 *Chronicle of the Second World War,* 648.

13 Martin Gilbert, 707.

14 *Ibid.,* 708.

[15] *Burning Conscience,* 99.

[16] *Ibid.,* 36.

[17] Ronnie Dugger, 174.

[18] *Ibid.,* 175.

[19] *Burning Conscience,* 25–26.

[20] *Ibid.,* 78.

[21] *Ibid.,* 82.

[22] Joseph B. Treaster, "Claude Eatherly, Hiroshima Spotter," *The New York Times,* Friday, July 7, 1978.

[23] Michael Bilton and Kevin Sim, *Four Hours in My Lai* (New York: Penguin, 1992), 24.

[24] *Ibid.,* 5–8.

[25] Personal communication, September 8, 1997.

[26] Dave Grossman, *On Killing: The Psychological Cost of Learning to Kill in War and Society* (Boston: Little, Brown & Co., 1996), 50.

[27] Steven Bentley, "VA Budget Cuts Add Insult to Injury," *Bangor (Maine) Daily News,* April 4–5, 1992.

Chapter 6

[1] Center for Defense Information, "Landmines: The Real Weapons of Mass Destruction," *The Defense Monitor,* vol. 25, no. 5 (July 1996).

[2] Anthony Lewis, "Suffer the Children," *The New York Times,* November 29, 1996.

[3] Zoltán Grossman, "One Hundred Years of Intervention," Committee Against Registration and the Draft, 1990.

[4] Iris Chang, "Exposing the Rape of Nanking," Book excerpt in *Newsweek,* December 1, 1997.

[5] Martin Gilbert, *The Second World War: A Complete History* (New York: Henry Holt, 1989), 746.

[6] *Ibid.,* 746.

[7] Iris Chang.

[8] James Trager, ed., *The People's Chronology: A Year-by-Year Record of Human Events from Prehistory to the Present* (New York: Holt, Rinehart & Winston, 1979), 982.

[9] Martin Gilbert, 746.

[10] James Trager, 982.

[11] *Ibid.*, 983.

[12] Martin Gilbert, 746.

[13] *Ibid.*, 745.

[14] Nhu T. Le, "Screaming Souls," *The Nation*, November 3, 1997.

[15] Zoltán Grossman.

[16] Zoltán Grossman, "Update to 'One Hundred Years of Intervention,'" Committee Against Registration and the Draft, 1995.

[17] Abdullah Matawi, "UN Sanctions on Iraq Lead to Deaths of 500,000 Children," *OneWorld News Service*, London, May 17, 1996.

[18] Center for Defense Information, "The Invisible Soldiers: Child Combatants," *The Defense Monitor*, vol. 26, no. 4, (July 1997).

[19] Tim O'Brien, *The Things They Carried* (New York: Penguin Books, 1990), 76, 84.

[20] Le Ly Hayslip with Jay Wurts, *When Heaven and Earth Changed Places: A Vietnamese Woman's Journey from War to Peace* (New York: Plume, 1990), xiv–xv.

[21] *Ibid.*, 326, 215, 70, 195–196, 200.

[22] Center for Defense Information, "Landmines."

[23] *Ibid.*

[24] Center for Defense Information, "Invisible Soldiers."

Chapter 8

[1] Jennifer Viereck, ed., *The Oil War Primer*, Nuremberg Actions, n.d., 37.

[2] Noam Chomsky, *What Uncle Sam Really Wants* (Tucson, Arizona: Odonian Press, 1996), 9–10.

[3] Center for Defense Information, "The Military and American Society: A Clash of Values," *The Defense Monitor,* vol. 22, no. 8, 1993.

[4] *Ibid.*

[5] Steven Bentley, "Agent Orange Still a Vital Concern to Vietnam Vets," *Perspective,* Lewiston, Maine, March 5, 1989.

[6] Steven Bentley, "In the Name of Freedom." *Bangor (Maine) Daily News,* July 9, 1991.

[7] Michael Bilton and Kevin Sim, *Four Hours in My Lai* (New York: Penguin, 1992), 256.

[8] *Ibid.,* 258.

[9] *Ibid.,* 44.

[10] *Ibid.,* 181.

[11] *Ibid.,* 310.

[12] *Ibid.,* 319.

[13] *Ibid.,* 315.

[14] *Ibid.,* 321.

[15] *Ibid.,* 320.

[16] *Ibid.,* 364.

[17] *Ibid.,* 204.

[18] Gregory L. Vistica, "A Quiet War Over the Past: The Brass Battles Over Giving a Hero of My Lai a Medal," *Newsweek,* November 24, 1997.

[19] Iris Chang, "Exposing the Rape of Nanking," Book excerpt in *Newsweek,* December 1, 1997.

[20] John Marks, "Swiss Cupidity, but German Evil," *U.S. News and World Report,* December 15, 1997.

[21] Martin Yant, *Desert Mirage: The True Story of the Gulf War* (Buffalo, New York: Prometheus Books, 1991), 209.

[22] *Ibid.,* 208–209.

[23] U.S. Congress, House, *Gulf War Veterans' Illnesses: VA, DOD Continue to Resist Strong Evidence Linking Toxic Causes to Chronic Health Effects,* 105th Congress, 1st Session, House Report 105–388, November 7, 1997.

Chapter 9

[1] 1 Chronicles 22:8

[2] "Exhibit Blunders Force Smithsonian Probe," *Air Force Magazine,* July 1995, 23.

[3] "An Interview with a Military Chaplain Who Served the Hiroshima and Nagasaki Bomb Squadrons," *Sojourners,* August 1980.

[4] George Zabelka, "Peace of Jesus Christ: Our Only Hope," Pax Christi 1985 (tape of speech obtained from Notre Dame University Archives, AO541).

[5] *Ibid.*

[6] Eberhard Arnold, "Love and Hatred in War Time," 1915, Bruderhof Archives, EA 15/16.

[7] Eberhard Arnold, "The Longing for Peace," 1915, Bruderhof Archives, EA 15/22.

[8] Eberhard Arnold, "The Influence of the World of the Spirits on Our Time," 1917, Bruderhof Archives, EA 17/18.

[9] Eberhard and Emmy Arnold, *Seeking for the Kingdom of God: Origins of the Bruderhof Communities* (Rifton, New York: Plough, 1974), 236, 237.

[10] Eberhard Arnold, *Inner Land: A Guide into the Heart and Soul of the Bible* (Rifton, New York: Plough, 1976), 270.

[11] Adolf Braun to Emil Becker, March 11, 1947.

[12] Congressional Record, June 16, 1934.

[13] Gloria Emerson, *Winners & Losers: Battles, Retreats, Gains, Losses, and Ruins from the Vietnam War* (New York: W. W. Norton, 1992), 210 (italics added).

[14] George A. Custer, *Wild Life on the Plains and Horrors of Indian Warfare* (North Stratford, New Hampshire: Ayer Company, 1980).

[15] Michael Bilton and Kevin Sim, *Four Hours in My Lai* (New York: Penguin, 1992), 3.

[16] *Ibid.,* 23.

[17] *Ibid.,* 367.

[18] *Ibid.,* 366.

[19] Thich Nhat Hanh, *Love in Action: Writings on Nonviolent Social Change* (Berkeley, California: Parallax Press, 1993), 73.

[20] *Ibid.*, 75–80.

Chapter 11

[1] Earthstewards Network, "PeaceTrees Vietnam: The Miracle Unfolds," Bainbridge Island, Washington, n.d.

[2] Chris Cowles, "Vets Set for Hecaling: Vietnam Bike Trek," Reuters, Hartford, Connecticut, December 26, 1997.

[3] "Vietnam Veteran Gets Letter from Son of Soldier He Killed," AP, Rochester, Illinois, *Kingston (New York) Sunday Freeman,* August 10, 1997.

[4] *Peace is Every Step: Meditation in Action: The Life and Work of Thich Nhat Hanh* (video), Mystic Fire Video, April 1998.

Chapter 12

[1] Thich Nhat Hanh, *Love in Action: Writings on Nonviolent Social Change* (Berkeley, California: Parallax Press, 1993), 88–89.

[2] Quoted in Aphrodite Matsakis, *Vietnam Wives: Facing the Challenges of Life with Veterans Suffering Post-Traumatic Stress* (Lutherville, Maryland: The Sidran Press, 1996), 3.

[3] The Bruderhof (the name means, literally, "place of brothers") is an international community committed to a life of non-violence and simplicity, based on Jesus' teachings. Ever since the first Bruderhof was founded in Germany, in 1920, community members have rejected private property, opting instead to pool not only their money and possessions, but their time and talents as well. Today, there are some 2500 adult members living at eight communities in the United States and Britain. At the heart of their commitment is a deep-seated dedication to service, family and love of neighbor.

Other Titles from Plough

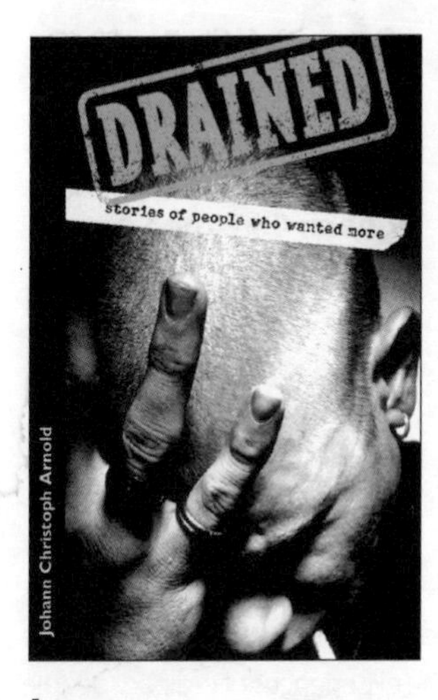

Drained

Stories of People Who Wanted More

Johann Christoph Arnold

152 pp., softcover, 0-87486-970-6
£5.50 / $8.00

IN *DRAINED,* ARNOLD TELLS the stories of people who overcame their greatest obstacles. He builds his book on the premise that the longing for peace lies deep within every person. *Drained* makes no attempt to present a cure-all for life's problems, but for those ready to go beyond quick-fix "remedies," it offers stepping stones to a fulfilled life.

A gem of a book…
Financial Times

A most interesting book…
Scotland on Sunday

Stories to sustain hope.
Desmond Tutu

The Lost Art of Forgiving

Stories of Healing from the Cancer of Bitterness

Johann Christoph Arnold
Foreword by Steve Chalke

149 pp., softcover, 0-87486-950-1
£7.99 / $13.00

THE LOST ART OF FORGIVING is a collection of stories showing the healing power of forgiveness in the experiences of ordinary people scarred by crime, betrayal, abuse, bigotry and war. Rather than offering a theoretical discussion, it lets the lives and voices of those who have forgiven, and those who haven't, speak for themselves.

A much-needed message for the whole world.
Nelson Mandela

Explains just how powerful the words "forgive me" can be. Nourishing…
Mail on Sunday

It's a fascinating book.
Lorraine Kelly, **TV and radio presenter**